# Visit us at

## www.syngress.com

Syngress is committed to publishing high-quality books for IT professionals and delivering those books in media and formats that fit the demands of our customers. We are also committed to extending the utility of the book you purchase via additional materials available from our Web site.

### SOLUTIONS WEB SITE
To register your book, visit www.syngress.com/solutions. Once registered, you can access our solutions@syngress.com Web pages. There you may find an assortment of valueadded features such as free e-books related to the topic of this book, URLs of related Web sites, FAQs from the book, corrections, and any updates from the author(s).

### ULTIMATE CDs
Our Ultimate CD product line offers our readers budget-conscious compilations of some of our best-selling backlist titles in Adobe PDF form. These CDs are the perfect way to extend your reference library on key topics pertaining to your area of expertise, including Cisco Engineering, Microsoft Windows System Administration, CyberCrime Investigation, Open Source Security, and Firewall Configuration, to name a few.

### DOWNLOADABLE E-BOOKS
For readers who can't wait for hard copy, we offer most of our titles in downloadable Adobe PDF form. These e-books are often available weeks before hard copies, and are priced affordably.

### SYNGRESS OUTLET
Our outlet store at syngress.com features overstocked, out-of-print, or slightly hurt books at significant savings.

### SITE LICENSING
Syngress has a well-established program for site licensing our e-books onto servers in corporations, educational institutions, and large organizations. Contact us at sales@syngress.com for more information.

### CUSTOM PUBLISHING
Many organizations welcome the ability to combine parts of multiple Syngress books, as well as their own content, into a single volume for their own internal use. Contact us at sales@syngress.com for more information.

SYNGRESS®

# Perl Scripting for Windows Security: Live Response, Forensic Analysis, and Monitoring

Harlan Carvey

| KEY | SERIAL NUMBER |
| --- | --- |
| 001 | HJIRTCV764 |
| 002 | PO9873D5FG |
| 003 | 829KM8NJH2 |
| 004 | BAL923457U |
| 005 | CVPLQ6WQ23 |
| 006 | VBP965T5T5 |
| 007 | HJJJ863WD3E |
| 008 | 2987GVTWMK |
| 009 | 629MP5SDJT |
| 010 | IMWQ295T6T |

PUBLISHED BY
Syngress Publishing, Inc.
Elsevier, Inc.
30 Corporate Drive
Burlington, MA 01803

Live Response, Forensic Analysis, and Monitoring

Printed and bound in the United Kingdom

Transferred to Digital Print 2011

ISBN 13: 978-1-59749-173-0

Publisher: Andrew Williams                    Page Layout and Art: SPi
Technical Editor: Dave kleiman               Copy Editor: Judy Eby

For information on rights, translations, and bulk sales, contact Matt Pedersen, Commercial Sales Director and Rights, at Syngress Publishing; email m.pedersen@syngress.com.

*To Terri and Kylie*

# Author

**Harlan Carvey** (CISSP), author of the acclaimed *Windows Forensics and Incident Recovery*, is a computer forensics and incident response consultant based out of the Northern VA/Metro DC area. He currently provides emergency incident response and computer forensic analysis services to clients throughout the U.S. His specialties include focusing specifically on the Windows 2000 and later platforms with regard to incident response, Registry and memory analysis, and post-mortem computer forensic analysis. Harlan's background includes positions as a consultant performing vulnerability assessments and penetration tests and as a full-time security engineer. He also has supported federal government agencies with incident response and computer forensic services.

Harlan holds a bachelor's degree in electrical engineering from the Virginia Military Institute and a master's degree in electrical engineering from the Naval Postgraduate School.

Harlan would like to thank his wife, Terri, for her support, patience, and humor throughout the entire process of writing his second book.

*Harlan wrote Parts I and II.*

# Technical Editor

**Dave Kleiman** (CAS, CCE, CIFI, CEECS, CISM, CISSP, ISSAP, ISSMP, MCSE, MVP) has worked in the Information Technology Security sector since 1990. Currently, he runs an independent Computer Forensic company DaveKleiman.com that specializes in litigation support, computer forensic investigations, incident response, and intrusion analysis. He developed a Windows Operating System lockdown tool, S-Lok, which surpasses NSA, NIST, and Microsoft Common Criteria Guidelines. He is frequently a speaker at many national security conferences and is a regular contributor to security-related newsletters, websites, and Internet forums. Dave is a member of many professional security organizations, including the Miami Electronic Crimes Task Force (MECTF), International Association of Computer Investigative Specialists (IACIS), International Information Systems Forensics Association (IISFA), the International Society of Forensic Computer Examiners (ISFCE), Information Systems Audit and Control Association (ISACA), High Technology Crime Investigation Association (HTCIA), Association of Certified Fraud Examiners (ACFE), High Tech Crime Consortium (HTCC), and the International Association of Counter Terrorism and Security Professionals (IACSP). He is also the Sector Chief for Information Technology at the FBI's InfraGard.

Dave was a contributing author for *Microsoft Log Parser Toolkit* (Syngress Publishing, ISBN: 1932266526), *Security Log Management: Identifying Patterns in the Chaos* (Syngress Publishing, ISBN: 1597490423) and, *How to Cheat at Windows System Administration* (Syngress Publishing ISBN: 1597491055). Technical Editor for *Perfect Passwords: Selection, Protection, Authentication* (Syngress Publishing, ISBN: 1597490415), *Winternals Defragmentation, Recovery, and Administration Field Guide* (Syngress Publishing, ISBN: 1597490792), *Windows Forensic Analysis: Including DVD Toolkit* (Syngress Publishing, ISBN: 159749156X), *The Official CHFI Study Guide* (Syngress Publishing, ISBN: 1597491977), and *CD and DVD Forensics* (Syngress Publishing, ISBN: 1597491284). He was Technical Reviewer for *Enemy at the Water Cooler: Real Life Stories of Insider Threats* (Syngress Publishing ISBN: 1597491292).

# Contributing Author

**Jeremy Faircloth** (Security+, CCNA, MCSE, MCP+I, A+, etc.) is an IT Manager for EchoStar Satellite L.L.C., where he and his team architect and maintain enterprisewide client/server and Web-based technologies. He also acts as a technical resource for other IT professionals, using his expertise to help others expand their knowledge. As a systems engineer with over 13 years of real-world IT experience, he has become an expert in many areas, including Web development, database administration, enterprise security, network design, and project management. Jeremy has contributed to several Syngress books, including *Microsoft Log Parser Toolkit* (Syngress, ISBN: 1932266526), *Managing and Securing a Cisco SWAN* (ISBN: 1932266917), *C# for Java Programmers* (ISBN: 193183654X), *Snort 2.0 Intrusion Detection* (ISBN: 1931836744), and *Security+ Study Guide & DVD Training System* (ISBN: 1931836728).

*Jeremy wrote Part III.*

# Contents

# Preface

## About the Book

I decided to write this book for a couple of reasons. One was that I've now written a couple of books that have to do with incident response and forensic analysis on Windows systems, and I used a lot of Perl in both books. Okay...I'll come clean...I used nothing but Perl in both books! What I've seen as a result of this is that many readers want to use the tools, but don't know how...they simply aren't familiar with Perl, with interpreted (or scripting) languages in general, and may not be entirely comfortable with running tools at the command line.

Another reason for writing this book is that contrary popular belief, there is no single application available that does everything or provides every function an incident responder could possibly need. By "popular", I'm primarily referring to those folks who don't perform incident response on a regular basis, as well as those who hire and have contracts with firms that provide incident responders and other consultants. Many times, incident responders (such as myself) will show up on-site will a pelican case full of equipment, CDs and DVDs full of tools and code, all of which provides a base capability. From there, what data to retrieve and how to view, manipulate, and present that data is dependant upon the customer...and no two are alike. In the years that I have been performing incident response and computer forensics, while I have had customers with similar requirements, no two engagements have been identical. Talking to other consultants, I have heard the same thing. There simply is no such thing as an application that will read Event Log file, web and FTP server log files, or perhaps entire images, and simply give you your answer (was the system compromised, by whom, and when) at the push of a button. Significant amounts of data collection, review, reduction, analysis, and presentation are required, and many times I find myself writing Perl scripts to perform one or more of those functions. In fact, I have found these scripts to be useful enough that for some, I have documented them, cleaned them up a bit, and provided them for public consumption.

I really need to point out that this book is *not* about computer forensic analysis. The purpose of this book is to show what can be (and has been) done, using Perl, to perform incident response,computer forensic analysis, and application monitoring on Windows systems. This book is about using Perl to complete computer incident response, forensic analysis tasks, and application monitoring, not about the tasks themselves, or the actual analysis.

# Who Should Read this Book

This book is intended for anyone who has an interest in useful Perl scripting, in particular on the Windows platform, for the purpose of incident response, and forensic analysis, and application monitoring. While a thorough grounding in scripting languages (or in Perl specifically) is not required, it helpful in fully and more completely understanding the material and code presented in this book. This book contains information that is useful to consultants who perform incident response and computer forensics, specifically as those activities pertain to MS Windows systems (Windows 2000, XP, 2003, and some Vista). My hope is that not only will consultants (such as myself) find this material valuable, but so will system administrators, law enforcement officers, and students in undergraduate and graduate programs focusing on computer forensics.

# Getting Started
## What is Perl?

Technically, Perl stands for "practical extraction and report language", and was originally developed as a general purpose programming language for manipulating text, but has grown into something much more. Perl is now used for a wide range of purposes, from automating system administration tasks, to use in web-based shopping carts, network- and web-development, etc.

Perl is an interpreted language, which means that once you've written your source code file, you don't need to compile the code into a standalone executable file, the way you do with other programming languages such as C or C++. Rather, you launch the interpreter, telling it to run your script, further passing any additional arguments that may be necessary. The interpreter checks and translates your code into something the operating system can use and understand, and then executes the commands in the script. This is a high-level view of things, of course, but my goal with this book isn't to teach you the philosophy of interpreted programming languages, but instead to give you something you can use.

Technical descriptions and the design of the programming language aside, Perl is a powerful tool for just about anyone involved with computers. Perl is extremely versatile, and can be used to perform a wide variety of tasks, some of which we'll be looking at in this book.

## Why use Perl?

Why use Perl? That's a great question.

One reason to use Perl is that it is fairly ubiquitous. There are a great number of platforms that have a version or distribution of Perl available. While our sole concern in this book is the Windows platform, Perl runs on Linux and Mac OS/X, as well as other platforms. What this means is that an examiner is not restricted to a specific platform on which to perform forensic analysis using Perl. With some care, Perl scripts can be written to run multiple platforms. I've written Perl scripts on a Windows system running on Intel hardware that ran equally well and produced identical output (given the same input file) on a Mac PowerPC system. This may be a concern where an examiner has a preference for her examination platform, or has some unique tools that are specific to that platform that she prefers to use for her analysis. Another concern may be when performing static analysis of Windows portable executable (PE) files or other potentially malicious code. On a Linux or Mac OS/X system, for example, the examiner won't suffer any ill effects if the executable file being examined is accidentally launched.

One of the major aspects of incident response and computer forensic analysis that I've seen is that no two incidents or investigations are alike. Even given nearly-identical computing infrastructures, different customers have different questions, based on their own concerns and the political make-up (i.e., personalities and goals of managers, etc.) of their organization. What this means is that when responding to an incident or performing forensic analysis, your tools may allow you to extract the raw data, but you're going to need some method of manipulating, correlating, and presenting that data in a manner that is required by the customer.

I've conducted examinations involving MS Outlook PST files, and where one examination required that I list the attachments by name, another required that I correlate emails and attachments found based on a keyword search against filenames within the acquired image that were founding during a search using the same list of keywords.

The point of this is that you're rarely going to find a commercial or freeware application that you can use during your examination, where all you have to do is click a button and the output will be exactly what you need, or (if you're a consultant) what your customer is asking for. Most available applications allow you to view the raw data in some form, and may assist you in doing a modicum of correlation, if any at all. Beyond that, however, it's up to the examiner to perform any additional correlation and presentation of the data that has been found. Sometimes this may require that the examiner translate binary data into something human-readable using a template or guide, or parsing through hundreds (or even thousands) of lines of log entries to extract those that are relevant, or perhaps correlate data between multiple files. Being able to produce a utility to perform this function in fairly short order can be of great benefit to an examiner as well as to her investigation.

Another example that comes to mind is running searches (for keywords, credit card numbers, social security numbers, etc.) across an acquired image and getting massive amounts of data, on the order of tens (or hundreds) of thousands of hits. These may need to be managed by filename path, credit card type, etc., and having to do this by hand can take several examiners days or even weeks to perform. However, with some programming ability, just-in-time utilities can be written to efficiently and accurately perform highly repetitive tasks, freeing the examiner to focus on other tasks.

As you can see, Perl has a number of advantages, but those advantages could apply to other languages, as well.

# How is Perl Used Within the Computer Security Community?

Perl is used extensively within the computer security community. (Not bad for an opening sentence, eh?)

The SleuthKit (http://www.sleuthkit.org) makes use of Perl. From the December 15, 2003 edition of The Sleuth Kit Informer:

*… it was originally designed to be a CGI script, so it was in one BIG Perl file …*

Further, the description for The Sleuthkit includes, " …The Sleuth Kit is written in C and Perl…".

The Metasploit Project (http://www.metasploit.org) makes use of Perl. HD Moore wrote the PEX, or Perl Exploit Library, a Perl module that "provides an object-oriented interface into common exploit development routines."

ProDiscover, the incident response and computer forensic analysis application from Technology Pathways (http://www.techpathways.com) uses Perl as its programming language. ProDiscover allows a forensic examiner to acquire images of systems, and then open those images for analysis. The ProDiscover graphical user interface (GUI) is fairly straightforward and intuitive, but Perl, implemented as ProScripts, can be used to automate tasks within the loaded project. The ProDiscover installation routine includes the ActiveState (http://www.activestate.com) ActivePerl distribution, as well as the ProScript.pm Perl module that provides the interface so that Perl can be used to interact with images loaded into ProDiscover projects. The Incident Response edition of ProDiscover also allows the responder to automate tasks such as distributing and connecting the PDServer agents, collecting volatile information, acquiring live images, and then disconnecting from the agent.

One of the reasons I use Perl in the work I do is that many times, there are no available tools that will do the work I need to do. I may be working on one investigation where I need to parse Registry files, and on the next one, I need to extract data from MS OutLook PST files. I've had multiple cases where I've had to parse PST files, but the requirements for each case was different; in one case, I had to simply obtain a list of file attachment names, whereas in another I had to correlate the list of attachment file names to the output of a keyword search. This work could be done by hand, but would take an inordinate amount of time. However, the point is that there are very often no available tools or applications that will allow you to do everything you may need to do; when performing forensic analysis, you may have no trouble obtaining the raw data, but that can often be thousands or even hundreds of thousands of entries, and the analysis of that data is the key to the work you need to do. Perl offers an excellent solution, in that code that you or someone else has previously written can be used to fill the gap quickly, and allow you to complete your work efficiently and more importantly, accurately.

# Getting Up and Running

## Installing Perl

The first thing you need to do in order to get started using Perl is to install a distribution for your platform. Perl has been ported to a number of platforms, as shown on the Ports page at the Comprehensive Perl Archive Network, or CPAN (http://www.cpan.org/ports). The Perl distribution used throughout this book is the ActivePerl distribution available from ActiveState. Once you've downloaded the most recent distribution of Perl, go ahead and install it. I usually install Perl into the "C:\Perl" directory, but you can install it into whichever directory you find most useful.

## Adding Modules

Perl ships with quite a number of installed modules. Modules are libraries of code that people have written that make repetitive tasks easier. Rather than constantly rewriting the code you use from scratch (say, to open sockets and connect to a server on the Internet) you can access the functionality you need in any one of a number of available modules. To see what modules were installed with Perl, you can click your way through the Start menu until you get to the ActivePerl Documentation page, which opens in your web browser.

Another way to manage Perl modules is to use the Perl Package Manager, or "ppm" that ships with ActivePerl. You access 'ppm' via the command line; simply open a command prompt, change directories to your Perl directory, and type "ppm /?" to get a list of commands you can use.

If you're not entirely comfortable with the command line, you can type "ppm" at the command prompt (with nothing else) and the ppm graphical user interface (GUI)[1] will open, as illustrated in Figure 1.

---

[1]  http://aspn.activestate.com/ASPN/docs/ActivePerl/5.8/faq/ActivePerl-faq2.html#ppm_gui

**Figure 1** PPM GUI (ppm-gui.tif)

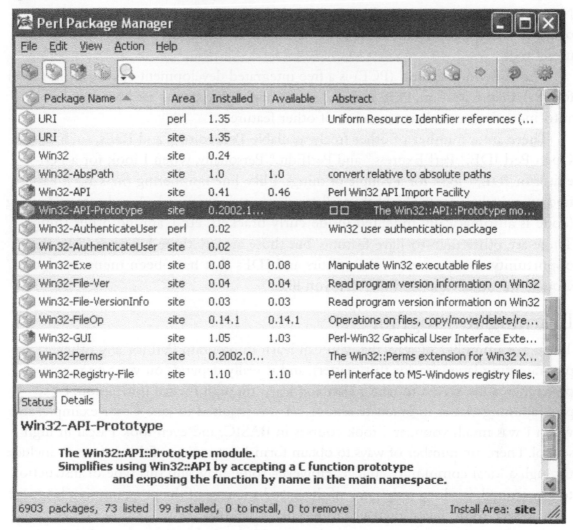

## Perl Editors

When writing Perl scripts, you need an editor of some kind. Back in my early days of graduate school (1994), those of us in the Electrical and Computer Engineering curriculum would write HTML pages using Notepad as our editor. You can use Notepad to write Perl scripts, as well, but I've found that using Notepad can make writing and troubleshooting Perl scripts a bit harder than it needs to be. When using an editor, the things I look for are syntax highlighting or color-coding, automating indenting (following curly brackets, etc.), and line numbering. These attributes make it easier to recognize my errors before I try running my code, and tracking them down when an error actually occurs.

There are a number of editors available for Perl. My personal favorite is UltraEdit.[2] Not only is UltraEdit an excellent Perl editor, but I use it to edit and view a variety of other formats, to include binary and hexadecimal. UltraEdit is a very versatile and useful tool.

The Perl Code Editor[3] (PCE) is a free integrated development environment (IDE) for Perl. Like UltraEdit, PCE includes syntax highlighting, line numbering, and auto-indenting, as well as a number of other features.

There are a number of other freely available Perl editors and IDEs, such as the Open Perl IDE,[4] Perl Express,[5] and PerlEdit.[6] Personally, when I look for a Perl editor or IDE, I look for a couple of things. I like line numbering (making it easy to find my mistakes), syntax highlighting (letting me catch my mistakes), and auto-indenting (code is automatically indented inside curly brackets, etc.), among other things. There are other nice-to-have features, but those are my three big ones. Take the opportunity to try some of the editors and IDEs that have been mentioned, or Google for others and find one that you like.

## Learning to Program

There are a number of ways that you can learn to program Perl (or any other programming language, for that matter), and it really depends on your own personal preference. One way is to take a class and learn through formal instruction. I had programming classes in graduate school ... I was required to take C, for example, and when I was much younger, I took courses in BASIC, and even took Pascal in high school. There are number of ways to obtain formal instruction of this nature, to include through a local community college. However, some may find this type of instruction too structured, teaching only some of the very basic uses of the programming language, such as how to do relatively trivial things like open files.

If you're so inclined, you can teach yourself, simply by diving in and doing it. There are a number of excellent resources available at of all places, your local library. By reading books and following the examples, you can learn to program quite quickly, picking up the basics before progressing on to more complex and useful tasks.

---

[2] http://www.ultraedit.com/
[3] http://www.perlvision.com/pce/
[4] http://open-perl-ide.sourceforge.net/
[5] http://www.perl-express.com/
[6] http://www.indigostar.com/perledit.html

An additional resource that is available is code that others have written. Some folks learn to program by looking at the steps others have taken to accomplish a task, and adding on to it, or modifying it in some other way to meet their needs. There are a number of resources available, through web sites, blogs, user forums, etc. There are number of resources that provide archives for code others have written and submitted, and there are folks out there who are willing to help, and provide assistance and advice (provided, of course, you're making an effort to perform the task yourself and not asking someone to do your homework for you).

## Writing Your Own Code

You'll see in the code throughout this book and on the accompanying DVD that I have my own programming style … there are certain ways that I do certain things in my code, and for me, that makes the code stand out. My hope is that it makes it easier for others to read and use, as well. Others have their own style, particularly in formatting. What's that joke about lawyers and opinions? Well, put five Perl programmers in a room with a task to accomplish, and as long as that task is beyond a simple "print" statement, you'll likely get five different versions of code as a result. Then, let them each look at the others and you'll likely get more. I mention this because I don't want you to think that my way of coding is THE way; it's simply A way. Many times, I will break certain tasks down into separate lines or sections of code, with documentation, where a single line may have been more elegant. I do this so that someone else, perhaps without as much background in either the problem or in Perl can then look at the code and have an easier time understanding what I did. There are also times where that "someone else" is me, six months or a year later. Sometimes elegance and speed have to give way to understandability and the ability to use the code again at a later date.

## Running Perl Scripts

Perhaps the biggest issue I have had with my first two books and Perl scripts is the inevitable emails that I get … "I double-clicked the Perl script and a black box flashed on the screen … what do I do?" Questions like this come from simply being (a) far too familiar and comfortable with GUI tools, and (b) unfamiliar with scripts of any kind (to include batch files) and the command prompt.

To run most Perl scripts, you need to open a command prompt, navigate to the appropriate directory, and then type in a command, by hand, finally hitting the Enter key. I know it sounds flippant, but I thought that perhaps breaking it down would make the process a bit easier to digest. In many cases, you may need to include parameters or arguments with the command; in essence, additional instructions which the script will process based on its code, and hopefully give you the desired result.

# Organization of the Book

## Part I

Part I addresses the use of Perl when working with live systems, as when an administrator is troubleshooting an issue, or when responding to an incident.

## Part II

Part II covers the use of Perl when performing forensic analysis of files after an image has been acquired of the system.

## Part III

In Part III we will be focusing on monitoring the core application processes, the core application dependencies, network connectivity, Web services, and log files.

## Download the Code

Visit www.syngress.com/solutions to download the Perl scripts from this book.

# Author Acknowledgements

I'd like to take this opportunity to acknowledge the efforts of a couple of folks who were instrumental to this book being written. First, I'd like to acknowledge God for blessing me, and my family for supporting me through the process of writing this book, as well as the others. I'd like to thank Dave Roth for his inspiration that started back in 1999, and for all of his assistance along the way. Dave provided support as I attempted to use his Perl modules, and even provided the drive to get me to present at my first conference. I'd like to thank Dave Schultz, whom I met while working for Trident Data Systems, for being patient as I fumbled, and for providing me with some useful programming hints that I still use today. I'd like to thank Jesse Kornblum, Andreas Schuster, and Didier Stevens for their drive and desire to push the envelope in the area of forensic analysis.

I'd like to thank the members of law enforcement who have asked for my help, and then acknowledged it. In a community that seems to harbor the expectation of free tools and tech support, it's a wonderful feeling when someone thanks you for your time and assistance.

There may be others that I'm missing, but I'd like to send out a heartfelt "thank you" to all those who chided (dare I say, "made fun of") me for using Perl in the first place… I know that some of you were kidding, while some of you were serious. Hopefully, folks that did both are reading these words.

# Perl Scripting and Live Response

## Solutions for this Part:

- Built-in Functions
- Running Processes
- Accessing the API
- WMI
- Accessing the Registry
- ProScripts

This Part focuses on the use of Perl when extracting data from a live system, as part of live response. "Live response" is a general term used to describe activities that are performed when information is needed from a system while it is still running. This most often involves collecting volatile data from a system, or data that is only available when the system is powered on and running. Live response activities can include something as simple as an administrator troubleshooting an issue on a system, or collecting process and network connection information from a system prior to powering the system down and acquiring an image of the system's hard drive. These activities can also include inventory control (determining who's logged into a system, what software is installed on a system, and so forth), and can be performed locally (while the administrator is sitting at the console) or remotely, over the network.

# Built-in Functions

ActiveState Perl comes with several built-in Windows (i.e., Win32) functions that allow you to access and retrieve specific information from a Windows system. For example, you can determine the current working directory (Win32::GetCwd()), the system architecture, and type of CPU of the system (Win32::GetArchName() and Win32::GetChipName(), respectively), as well as a number of other very useful pieces of information. All of these functions are simply interfaces into the appropriate Windows application program interface (API) function calls, and allow the programmer to quickly retrieve the information they're looking for.

## Win32.pl

Demonstrates the use of some of the Perl Win32 built-in functions:

```
use strict;
use Win32;
print "Architecture  : ".Win32::GetArchName()."\n";
print "Chip          : ".Win32::GetChipName()."\n";
print "Perl Build    : ".Win32::BuildNumber()."\n";
print "Node Name     : ".Win32::NodeName()."\n";
print "Login Name    : ".Win32::LoginName()."\n";
print "OS Name       : ".Win32::GetOSName()."\n";
my ($str,$maj,$min,$build,$id) = Win32::GetOSVersion();
print "$str $maj $min $build $id\n";
```

On my test system the output from this script appears as follows:

```
C:\Perl>win32.pl
Architecture  : x86
Chip          : 586
Perl Build    : 819
Node Name     : WINTERMUTE
Login Name    : Harlan
OS Name       : WinXP/.Net
Service Pack 2 5 1 2600 2
```

As you can see, some of this information can be quite useful during incident response. Check the ActiveState Perl documentation for a complete list of Win32 functions.

# Pclip.pl

While not a built-in function, ActiveState Perl ships with several Perl modules that are specific to the Windows platform. For example, the Win32::Clipboard module allows you to set or retrieve the contents of the Windows Clipboard.

```
use strict;
use Win32::Clipboard;
my $clip = Win32::Clipboard();
my $clipboard;
if ($clipboard = $clip->Get()) {
  print "Clipboard Contents\n";
  print "-" X 20,"\n";
  print $clipboard."\n";
}
else {
  print "Error retrieving clipboad contents:
".Win32::FormatMessage(Win32::GetLastError())."\n";
}
```

Many times during incident response, there may be information available on the clipboard that may be of use to the investigator, such as portions of an e-mail or document, a password, or text transferred between windows on the desktop. The Win32::Clipboard module allows you to retrieve the contents of the clipboard, and display it in any way that is useful to you. Pclip.pl is a very simple example of the use of the module. Consult the Perl "plain old documentation" (POD) for the module for some ideas of a more complete script that is capable of handling bitmaps, lists of files, or other data formats.

As an example, I was looking up some directions to a location that I needed to visit, and that I had to provide to a friend. I found the street address of the location and selected it in one Web page window, copied it, and pasted it into the e-mail that I was preparing to send. Afterward, I ran pclip.pl and this is what I got back:

```
Clipboard Contents
-------------------
123 Fake Street
```

Imagine what people copy into their clipboards throughout the day, many without really understanding what happens. I suggest that just as an experiment, you should go around an office or a school, or you can even do this at home, and simply open a Notepad window, place the cursor anywhere within the window, and press Ctrl-V. Whatever is in the clipboard will be pasted into Notepad. Pclip.pl allows you to automate this collection process.

# Running Processes

When performing live response, we are working with and interacting with a live, running system. Many times, when responding to an incident, a user may still be logged into the system. In some cases, such as employee workstations within an organization, this user may be the employee themselves. In others, such as in server rooms and data centers, this user will most likely be a system administrator. Often, an incident will occur and we will need to log into the system ourselves (as a consultant, I always have the system administrator do that) in order to obtain information from a system. The point is that in order to collect information from a live system, there has to be an account logged into the system, either at the console (via the keyboard) or over the network.

As the system is live and running, there are processes running, threads being executed, and code being processed. This is how we interact with the system; we "ask" the system for information by running processes ourselves. Our Perl scripts may be processes, but many times it is simply much easier to run external, third-party tools, or even tools that are native to the system itself, in order to get the information we need. For example, let's say that we'd like to get a list of open network connections from a system. The first thing that comes to mind as a means of requesting this information from the system is the native utility, netstat.exe.

One question that may immediately come to mind is, if I can run netstat.exe (or any other tool) from the command line, why bother to do it via a Perl script? Well, there are a couple of very good answers to that. One is that by including the use of the tool or utility in a Perl script (or batch file), we have a form of self-documentation. Documentation is a very important aspect of incident response. Second, many of the tools we may want to run on systems have a number of command-line arguments, and I don't know about you, but sometimes in the heat of the moment, I may not be able to keep that information straight, particularly at 2:30 A.M. when I'm trying to collect information from systems' that may have been compromised. So, by including the tool or utility in a Perl script, I have a degree of automation that prevents me from making mistakes, particularly through repetition. Finally, it's not often that I deal with only one system, or one tool or utility. Most often, I'm responding to 10, 50, or 100 systems, and I'm running a number of different tools on each of those systems. Using a Perl script, I'm able to put everything into a single command so that when the situation changes, I'm prepared. That way, if something happens further down the road and someone asks me what I did, I can refer back to the Perl script and the copies of the tools I ran.

So, there are a couple of ways that we can run programs on a system. Using netstat.exe, we'll take a look at several of them. Do not think that these are the only ways to address this particular issue. One of the strengths of Perl is that there is usually more than one way to complete a task. What I'm going to do here is show you some of what I have come up with, but this does not mean that these methods or Perl scripts are the *only* way to do things.

# Netstat1.pl

Perhaps the simplest way to launch external programs in Perl is to use the system() function. The system() function simply forks a child process from a parent process, which waits for the child process to complete, and then exits. A very simple use of the system() function, using netstat.exe as our example, is as follows:

```
use strict;
my @args = ("netstat", "-ano");
system(@args);
```

While this could have been much simpler in only a single line, simplicity or elegance isn't the issue here. What happens when we run this code is that the output of our command appears at the console, or standard output (i.e., STDOUT). So all we've really done here is added a layer of abstraction and not really bought ourselves anything useful. In order to save the output of the command, for example, we'd still need to use the redirection operator at the command prompt:

```
C:\>perl nestat1.pl > netstat.log
```

That's really no different from not using Perl at all:

```
C:\>netstat -ano > netstat.log
```

So, a bit of extra effort, but it would appear that we really haven't bought ourselves anything. Now, this might be different if we were using this script to run multiple commands; after all, wouldn't we then be benefiting from automation? During incident response, you're usually under pressure, either from your boss or the clock, or you're tired because it's 3:00 A.M., and the first thing that will happen is that you'll forget a command or mistype a command or something that will be frustrating under those conditions. By linking the commands together into a script, we can now type in a single command, a couple of short keystrokes, and have everything run for us. However, at this point, we really don't have anything much more than a batch file contained in a Perl script. We haven't taken full advantage of the power of Perl to make our jobs easier.

# Netstat2.pl

Another way to run external commands through Perl is to use backticks. Backticks are not the single quote operator on your keyboard; rather the backticks are the slanted single quote operator. Using the backtick operator, you can access system commands or even external commands (replace netstat.exe with your program of choice, ensuring that it is located in the PATH). For example, let's call the following code "netstat2.pl":

```
#! c:\perl\bin\perl.exe
use strict;
my @netstat = `netstat -ano`;
map{print "$_"}@netstat;
```

Notice that we launch netstat.exe with the "a," "n," and "o" switches, and collect the output of the command, whatever it may be, into a Perl list (or array). From there, the script finishes by simply printing out what's in the list. Now, the output of the command isn't at the console (STDOUT); rather, we've got control of that output and we can do what we like with it.

## Master Craftsman

### Extending the Use of Backticks

You can use the backticks to not only launch applications on the system, or even applications and programs external (i.e., not native) to the system, but also to access native commands, such as "dir." "Dir" doesn't exist as an executable file on a system, but it is an accessible command.

Other things you can do is include a list of commands in an array (such as dir /ah, netstat –ano, and so forth) and then iterate through the list, running each command individually. If you're interested in running several commands and correlating the output or filtering the output, the Perl lists make that very easy to do.

Now we're at the point where we're making our jobs a little easier. For example, I can filter through the output, looking for a particular Internet Protocol (IP) address, or skipping lines that contain the loopback address (127.0.0.1). I can minimize the output, showing only the things I want to see, rather than showing me everything. I can filter the data, showing only those network connections that are in a particular state, such as LISTENING, TIME_WAIT, or ESTABLISHED. The point is, we're now making our jobs easier by running a command of our choosing and being able to manage the output of that command.

# Netstat3.pl

The Win32::Job module provides a bit more granularity of control when creating and running processes, as shown in netstat3.pl below.

```
#! c:\perl\bin\perl.exe
use strict;
use Win32::Job;
eval {
      my $job = Win32::Job->new();
      my $result = $job->spawn("netstat.exe","netstat.exe -ano");
      die "Value is undefined. ".$^E."\n" unless (defined $result);
      my $ok = $job->run(60);
};
print $@."\n" if ($@);
```

When we run netstat3.pl, we get the same sort of output we would expect to see if we were running the netstat –ano command from the command line; however, in this case, we are able to use the Win32::Job module to do things such as limit the amount of time that the process runs. In netstat3.pl, we limit that time to 60 seconds, which is a long time, and probably more time than we need in most cases. However, I have seen simple command-line tools (such as netstat.exe) hang when run on some systems, or simply take an inordinate amount of time to run (due to high processing overhead from other processes, and so forth). In such cases, we may want to limit how long the process runs, and that's where Win32::Job comes in.

There are a couple of other functions within the Win32::Job module that may be of use, depending upon what you're doing and the level of control of the process you wish to achieve. For example, you can use the spawn() function to redirect STDOUT and STDERR messages to log files, or you can use the watch() function to provide a handler for the process, in order to achieve an even more granular level of control over the process. Check the POD for the Win32::Job module and for the Win32:: Process module, for other ideas on how to run external processes from within Perl code.

Also notice the use of the eval{} block. This allows us to tell Perl to evaluate the code, and trap any errors that may occur. One of the big ones that occurred when I was writing and testing the above code was that I had misspelled the name of the executable (i.e., "nestat.exe" instead of "netstat.exe"). While this is not an error that would cause a major application crash, the error was trapped, nonetheless. The eval{} block is useful for trapping such errors, and even allowing your code to progress in the event of an error that you simply wish to recover from (and not have your entire script bomb out!).

# Accessing the API

When performing live response or perhaps even analyzing files retrieved from a system during live response, you may want to access the Windows API. The Windows API can provide some useful functionality, already partially built. Fortunately, Microsoft exposes a good portion of the public API via the Microsoft.com Web site, and in addition, there are books available that describe other API functions that are accessible, albeit not fully documented.

In order to access the Windows API, you need to be sure that you have the Win32::API module installed. You can check to see if this module has been installed in your Perl distribution by typing the following command at the command prompt.

```
C:\perl>ppm query Win32-api
```

Figure I.1 illustrates the output of this command on my system.

**Figure I.1** Querying for the Win32::API Module

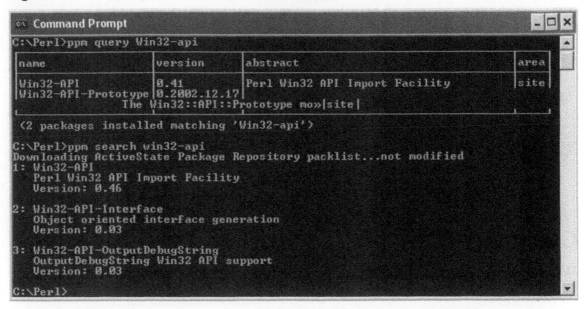

You'll notice in Figure I.1 that when I ran my query for "win32-api," all of the modules that began with that name were returned. What is this module named "Win32-API-Prototype[1]"? This is a module created by Dave Roth that encapsulates the Win32::API module and makes the Win32::API module easier to use.

---

[1] www.roth.net/perl/prototype/

**TIP**

To install the Win32::API::Prototype module, type the following command:
    **C: \perl>ppm install http://www.roth.net/perl/packages/win32-api-prototype.ppd**

**NOTE**

By "easier to use" on the Web page that describes the Win32::API::Prototype module, Dave provides several examples of how to use a module to access API functions, as well as how to set up and format the various arguments. Dave uses a list (array) called "@ParameterTypes" to describe and hold the various types of the parameters or arguments of the function.

# Getsys.pl

When performing incident response, some of the information you may want to get from the live system includes things such as the current system time and the uptime of the system. All of these pieces of information can be retrieved via the Windows API. To do so, we'll use Dave's Win32::API::Prototype module to access a couple of Windows API calls:

```perl
#! c:\perl\bin\perl.exe
#-----------------------------------------------------------------------------
# getsys.pl
# This script demonstrates the use of the Win32::API::Prototype
# module to retrieve time-based information from the local system
#
# The only required module is Win32::API::Prototype:
# ppm install http://www.roth.net/perl/packages/win32-api-prototype.ppd
#
# Usage: [perl] getsys.pl
#
# Copyright 2001-2007 H. Carvey
#-----------------------------------------------------------------------------
use strict;
use Win32::API::Prototype;
my @month = qw/Jan Feb Mar Apr May Jun Jul Aug Sep Oct Nov Dec/;
my @day = qw/Sun Mon Tue Wed Thu Fri Sat/;
```

```perl
# Meanings of the following constants can be found here:
# http://msdn.microsoft.com/library/default.asp?url
#     =/library/en-
#     us/sysinfo/base/gettimezoneinformation.asp
my @tz = qw/TIME_ZONE_ID_UNKNOWN TIME_ZONE_ID_STANDARD TIME_ZONE_ID_DAYLIGHT/;
ApiLink('kernel32.dll',
    'VOID GetSystemTime(LPSYSTEMTIME lpSystemTime)')
  || die "Cannot locate GetSystemTime()";
ApiLink('kernel32.dll',
    'DWORD GetTimeZoneInformation(
    LPTIME_ZONE_INFORMATION lpTimeZoneInformation)')
    || die "Cannot locate GetTimeZoneInformation()";
# The return value is the number of milliseconds that
#      have elapsed since the system was started.
# This value rolls over to zero after 49.7 days
ApiLink('kernel32.dll',
    'DWORD GetTickCount()')
    || die "Cannot locate GetTickCount()";
# Get the system time
# Ref: http://msdn.microsoft.com/library/default.asp?url=
#      /library/en-us/sysinfo/base/getsystemtime.asp
my $lpSystemTime = pack("S8", 0);
GetSystemTime($lpSystemTime);
my $str = sys_STR($lpSystemTime);
print "System Time : $str\n";
my ($day,$hour,$min,$sec) = getUpTime();
print "System Uptime: $day days, $hour hours, $min min, $sec sec.\n";
print "\n";
my $lpTimeZoneInformation = pack '1A64SSSSSSSS1A64SSSSSSSS1',
0, ' ' X 64, 0, 0, 0, 0, 0, 0, 0, 0, 0, ' ' X 64, 0, 0, 0, 0, 0, 0, 0, 0, 0;
my $bias;
my $standardName;
my $standardBias;
my $dayLightName;
my $dayLightBias;
my @c;
my @f;
my $ret = GetTimeZoneInformation($lpTimeZoneInformation);
($bias, $standardName, $c[0], $c[1], $c[2], $c[3], $c[4], $c[5], $c[6], $c[7],
    $standardBias, $dayLightName, $f[0], $f[1], $f[2], $f[3], $f[4], $f[5],
$f[6], $f[7],
```

```perl
    $dayLightBias) = unpack '1A64SSSSSSSS1A64SSSSSSSS1', $lpTimeZoneInformation;
print "Return code => ".$tz[$ret]."\n";
# The bias is the difference, in minutes, between UTC time and local time.
# Convert to hours for presentation
# UTC = local time + bias
print "Bias   => ".$bias." minutes\n";
if (1 == $ret) {
  print "Standard Bias => ".$standardBias." minutes\n";
}
elsif (2 == $ret) {
  print "Daylight Bias => ".$dayLightBias." minutes\n";
}
else {
# do nothing
}
$standardName =~ s/\00//g;
$dayLightName =~ s/\00//g;
print "StandardName => ".$standardName."\n";
print "DaylightName => ".$dayLightName."\n";
# Convert returned SystemTime into a string
sub sys_STR {
   my $lpSystemTime = $_[0];
   my @time = unpack("S8", $lpSystemTime);
  $time[5] = "0".$time[5] if ($time[5] =~ m/^\d$/);
  $time[6] = "0".$time[6] if ($time[6] =~ m/^\d$/);
  my $timestr = $day[$time[2]]." ".$month[$time[1]-1]." ".
    $time[3]." ".$time[4].":".$time[5].":".$time[6]." ".
    $time[0];
  return "$timestr";
}
sub getUpTime {
   my $count = GetTickCount();
   my $sec = 1000;
   my $min = $sec * 60;
   my $hour = $min * 60;
   my $day = $hour * 24;
   my ($temp,$d,$h,$m,$s);
   if ($count > $day) {
     $d = (split(/\./,$count/$day,2))[0];
     $temp = $count%$day;
```

```perl
      $h = (split(/\./,($temp/$hour),2))[0];
      $temp = $temp%$hour;
      $m = (split(/\./,($temp/$min),2))[0];
      $temp = $temp%$min;
      $s = (split(/\./,($temp/$sec),2))[0];
   }
   elsif ($count > $hour) {
     $d = 0;
     $h = (split(/\./,($count/$hour),2))[0];
     $temp = $count%$hour;
     $m = (split(/\./,($temp/$min),2))[0];
     $temp = $temp%$min;
     $s = (split(/\./,($temp/$sec),2))[0];
   }
   elsif ($count > $min) {
     $d = 0;
     $h = 0;
     $m = (split(/\./,($count/$min),2))[0];
     $temp = $count%$min;
     $s = (split(/\./,($temp/$sec),2))[0];
   }
   elsif ($count > $sec) {
     $d = 0;
     $h = 0;
     $m = 0;
     $s = (split(/\./,($count/$sec),2))[0];
   }
   return ($d,$h,$m,$s);
}
```

Running the script returns the following information from my system:

```
C:\Perl>getsys.pl
System Time  : Fri Aug 31 22:57:38 2007
System Uptime: 0 days, 12 hours, 6min, 35sec.
Return code        => TIME_ZONE_ID_DAYLIGHT
Bias               => 300 minutes
Daylight Bias      => -60 minutes
StandardName       => Eastern Standard Time
DaylightName       => Eastern Daylight Time
```

## Master Craftsman

### Getting Even More Information

You can extend the getsys.pl script to get things such as the current system time, the current Universal Coordinated Time (UTC) (UTC is analogous to Greenwich Mean Time [GMT]), the system name, the name of the logged on user, and so forth. For example, to get the system name, you might use the GetComputerNameA[2] API function, and to get the name of the logged on user, you might use the GetUserNameA[3] API function.

Retrieving information from a Windows system via the API can be useful, but it can also lead to problems. Many times, APIs will change between versions of Windows (such as between Windows 2000 and XP), or they may even change when a Service Pack is installed or updated. As such, direct use of the Windows API to collect some information from systems should be thoroughly tested before being deployed on a widespread basis.

# WMI

The Windows Management Instrumentation (WMI) is a great way to obtain information from live Windows systems. WMI is really nothing more than many of the hard-core details of accessing the Windows API that have been encapsulated and made easier to use. Instead of having to write code that accesses a system to determine what version of Windows it is and then take appropriate steps based on that version, an administrator can write code that will work (in most cases) consistently across Windows 2000 all the way through Vista. This means that an administrator or incident responder can request a list of the active processes from systems from across the enterprise, either locally on the host systems or remotely from a centrally located management console, and use the same code to get the same results, regardless of the version of Windows being queried. The advantage of this is that during incident response, many times some tools work better on some systems than on others and some tools simply do not work at all.

---

[2] http://msdn2.microsoft.com/en-us/library/ms724295.aspx
[3] http://msdn2.microsoft.com/en-us/library/ms724432.aspx

Another advantage of WMI is that it provides a cleaner, easier to use interface to some (albeit not all) of what you can access via the Win32::API and Win32::API:: Prototype modules. For example, you can access information about the microprocessor, physical memory, hard drives, and other devices on the systems.

The Win32::OLE module provides the interface through which you can use Perl to access the WMI classes. The WMI classes provide access to operating system classes[4], such as classes that provide access to information pertaining to files, processes, drivers, networking, operating system settings, and so forth. The computer system hardware classes[5] provide access to information about devices on the system, such as the processor(s), hard drivers, batteries, fans, and so forth.

# Fw.pl

While one advantage of the WMI classes is that they provide a common interface to certain aspects of the Windows platform regardless of the operating system version, one disadvantage is that some versions of Windows have functionality that others do not. For example, Windows XP Service Pack 2 and Windows 2003 have a built-in firewall that is part of the Security Center, something neither Windows NT 4.0 (WMI classes were installed as a separate download for Windows NT) nor Windows 2000 have.

```
#! c:\perl\bin\perl.exe
#-------------------------------------------------------------------------------
# fw.pl
# Use WMI to get info about the Windows firewall, as well as
# information from the SecurityCenter
#
# Usage: fw.pl [-bsph] [-app] [-sec]
#
# copyright 2006-2007 H. Carvey keydet89@yahoo.com
#-------------------------------------------------------------------------------
use strict;
use Win32::OLE qw(in);
use Getopt::Long;
my %config = ();
```

---

[4] http://msdn2.microsoft.com/en-us/library/aa392727.aspx
[5] http://msdn.microsoft.com/library/default.asp?url=/library/en-us/wmisdk/wmi/
computer_system_hardware_classes.asp

```perl
Getopt::Long::Configure("prefix_pattern=(-|\/)");
GetOptions(\%config, qw(b s sec p app help|?|h));
# if -h, print syntax info and exit
if ($config{help}) {
    \_syntax();
    exit 1;
}
# some global hashes used throughout the code
my %proto =     (6 => "TCP",
                 17 => "UDP");
my %ipver =     (0 => "IPv4",
                 1 => "IPv6",
                 2 => "Any");
my %type =      (0 => "DomainProfile",
                 1 => "StandardProfile");
print "[".localtime(time)."] Checking Windows Firewall on ".Win32::NodeName()."...\n"
  unless ($config{sec});
# Create necessary objects
my $fwmgr = Win32::OLE->new("HNetCfg.FwMgr")
    || die "Could not create firewall mgr obj: ".Win32::OLE::LastError()."\n";
my $fwprof = $fwmgr->LocalPolicy->{CurrentProfile};
if (! %config || $config{b}) {
# Profile type: 0 = Domain, 1 = Standard
    print "Current Profile = ".$type{$fwmgr->{CurrentProfileType}}." ";
    if ($fwprof->{FirewallEnabled}) {
      print "(Enabled)\n";
    }
    else {
      print "(Disabled)\n";
      exit(1);
    }
    ($fwprof->{ExceptionsNotAllowed}) ?(print "Exceptions not allowed\n"):
(print "Exceptions allowed\n");
    ($fwprof->{NotificationsDisabled})?(print "Notifications Disabled\n"):
(print "Notifications not disabled\n");
    ($fwprof->{RemoteAdminSettings}->{Enabled}) ? (print "Remote Admin Enabled\n") :
(print "Remote Admin Disabled\n");
    print "\n";
}
if (! %config || $config{app}) {
    print "[Authorized Applications]\n";
```

```perl
    foreach my $app (in $fwprof->{AuthorizedApplications}) {
      if ($app->{Enabled} == 1) {
        print $app->{Name}." - ".$app->{ProcessImageFileName}."\n";
        print "IP Version = ".$ipver{$app->{IPVersion}}."; Remote Addrs = "
.$app->{RemoteAddresses}."\n";
        print "\n";
      }
    }
}
if (! %config || $config{p}) {
    print "[Globablly Open Ports]\n";
    foreach my $port (in $fwprof->{GloballyOpenPorts}) {
      if ($port->{Enabled} == 1) {
        my $pp = $port->{Port}."/".$proto{$port->{Protocol}};
        printf "%-8s %-35s %-20s\n",$pp,$port->{Name},$port->{RemoteAddresses};
      }
    }
    print "\n";
}
if (! %config || $config{s}) {
    print "[Services]\n";
    foreach my $srv (in $fwprof->{Services}) {
      if ($srv->{Enabled}) {
        print $srv->{Name}." (".$srv->{RemoteAddresses}.")\n";
        foreach my $port (in $srv->{GloballyOpenPorts}) {
          if ($port->{Enabled} == 1) {
            my $pp = $port->{Port}."/".$proto{$port->{Protocol}};
            printf " %-8s %-35s %-20s\n",$pp,$port->{Name},$port->{RemoteAddresses};
          }
        }
        print "\n";
      }
    }
}
# Check the SecurityCenter for additional, installed, WMI-managed FW and/or
AV software
# Some AV products are not WMI-aware, and may need a patch installed
if ($config{sec}) {
    my $server = Win32::NodeName();
    print "[".localtime(time)."] Checking SecurityCenter on $server...\n";
```

```
   my $objWMIService = Win32::OLE->GetObject("winmgmts:\\\\$server\\root\\
SecurityCenter") || die "WMI connection failed.\n";
# Alternative method
# my $locatorObj = Win32::OLE->new('WbemScripting.SWbemLocator') || die
#       "Error creating locator object: ".Win32::OLE->LastError()."\n";
# $locatorObj->{Security_}->{impersonationlevel} = 3;
# my $objWMIService = $locatorObj->ConnectServer($server,'root\
SecurityCenter',"","")
#       || die "Error connecting to $server: ".Win32::OLE->LastError()."\n";
   my $fwObj = $objWMIService->InstancesOf("FirewallProduct");
   if (scalar(in $fwObj) > 0) {
     foreach my $fw (in $fwObj) {
       print "Company = ".$fw->{CompanyName}."\n";
       print "Name = ".$fw->{DisplayName}."\n";
       print "Enabled = ".$fw->{enabled}."\n";
       print "Version = ".$fw->{versionNumber}."\n";
     }
   }
   else {
     print "There do not seem to be any non-MS, WMI-enabled FW products
installed.\n";
   }
   my $avObj = $objWMIService->InstancesOf("AntiVirusProduct");
   if (scalar(in $avObj) > 0) {
     foreach my $av (in $avObj) {
       print "Company = ".$av->{CompanyName}."\n";
       print "Name = ".$av->{DisplayName}."\n";
       print "Version = ".$av->{versionNumber}."\n";
       print "O/A Scan = ".$av->{onAccessScanningEnabled}."\n";
       print "UpToDate = ".$av->{productUptoDate}."\n";
     }
   }
   else {
     print "There do not seem to be any WMI-managed A/V products installed.\n";
   }
}
sub _syntax {
   print>> "EOT";
fw [-bsph] [-app]
Collect information about the Windows firewall (local system only) and
the SecurityCenter (additional WMI-managed FW and AV products)
```

```
 -b .........Basic info about Windows firewall only
 -app .......Display authorized application info for the Windows firewall
(enabled only)
 -s .........Display service info for the Windows firewall (enabled only)
 -p .........Display port info for Windows firewall (enabled only)
 -sec .......Display info from the SecurityCenter (other installed,WMI-
     managed FW and/or AV)
 -h .........Help (print this information)
Ex: C:\\>fw -s >server> -u >username> -p >password>
copyright 2006-2007 H. Carvey
EOT
}
```

There are a couple of things you'll notice about the fw.pl Perl script. One is the use of the Getopt::Long module in order to allow for the use of command-line arguments in the script. This allows us to program different functionality into a single script, rather than writing separate scripts to do slightly different things. For example, if you look at the content of the _syntax() function from the script, you'll see that you can use command-line arguments and switches to modify the output of the script and show different bits of information. This way, we can have one script with a complete set of functionality, rather than half a dozen different scripts. If I run the fw.pl script on my own system with just the "-b" switch, I get the following output:

```
C:\Perl>fw.pl -b
[Fri Aug 31 17:29:47 2007] Checking Windows Firewall on WINTERMUTE...
Current Profile = StandardProfile (Enabled)
Exceptions allowed
Notifications not disabled
Remote Admin Disabled
```

Running the script with just the "-s" switch to see the service information for the firewall, I get:

```
C:\Perl>fw.pl -s
[Fri Aug 31 17:31:41 2007] Checking Windows Firewall on WINTERMUTE...
[Services]
File and Printer Sharing (LocalSubNet)
  139/TCP NetBIOS Session Service   LocalSubNet
  445/TCP SMB over TCP              LocalSubNet
  137/UDP NetBIOS Name Service      LocalSubNet
  138/UDP NetBIOS Datagram Service LocalSubNet
```

```
UPnP Framework (LocalSubNet)
  1900/UDP SSDP Component of UPnP Framework       LocalSubNet
  2869/TCP UPnP Framework over TCP LocalSubNet
```

Using just the "-sec" switch to check the SecurityCenter[6] settings, I get:

```
C:\Perl>fw.pl -sec
[Fri Aug 31 17:32:57 2007] Checking SecurityCenter on WINTERMUTE...
There do not seem to be any non-MS, WMI-enabled FW products installed.
There do not seem to be any WMI-managed A/V products installed.
```

Now, had I had a WMI-enabled antivirus product installed on this system, it would show up in the output of the script. I would also be able to get some setting information from the system regarding an installed WMI-enabled firewall, if there is one, as in the following output taken from another system:

```
D:\Programs\Perl>fw.pl -sec
[Thu Sep 6 15:23:15 2007] Checking SecurityCenter on A1...
Company   = Check Point, LTD.
Name      = ZoneAlarm Pro Firewall
Enabled   = 1
Version   = 7.0.337.000
Company   = GRISOFT
Name      = AVG 7.5.485
Version   = 7.5.485
O/A Scan  = 1
UpToDate  = 1
```

As you can see, this system has the ZoneAlarm Pro Firewall and Grisoft AVG anti-virus (AV) applications installed (and more importantly, enabled), and the AV product appears to be up-to-date and enabled.

# Nic.pl

The Perl script nic.pl allows you to retrieve information about network interface cards (NICs) through the Win32_NetworkAdapterConfiguration[7] WMI class. The script allows you to collect information from either the local system, or from a remote system. While the class, like many other WMI classes, provides functions or methods for modifying information on the system, during incident response we're most interested in collecting information, so we'll stick to simply querying the system and retrieving the information that we need, and avoid modifying anything.

---

[6] http://support.microsoft.com/kb/883792
[7] http://msdn2.microsoft.com/en-us/library/aa394217.aspx

```perl
#! c:\perl\bin\perl.exe
#------------------------------------------------------------------
# nic.pl
# Use WMI to get information about active network interface cards
# on a system
#
# Usage: [perl] nic.pl
#
# Copyright 2004-2007 H. Carvey keydet89@yahoo.com
#------------------------------------------------------------------
use strict;
use Win32::OLE qw(in);
use Getopt::Long;
my %config = ();
Getopt::Long::Configure("prefix_pattern=(-|\/)");
GetOptions(\%config, qw(server|s=s user|u=s passwd|p=s csv|c help|?|h));
if ($config{help}) {
    \_syntax();
    exit 1;
}
if (! %config) {
    $config{server} = Win32::NodeName();
    $config{user} = "";
    $config{passwd} = "";
}
$config{user} = "" unless ($config{user});
$config{passwd} = "" unless ($config{passwd});
my $locatorObj = Win32::OLE->new('WbemScripting.SWbemLocator') || die
    "Error creating locator object: ".Win32::OLE->LastError()."\n";
$locatorObj->{Security_}->{impersonationlevel} = 3;
my $serverObj = $locatorObj->ConnectServer($config{server},'root\cimv2',$config{user}
,$config{passwd})
    || die "Error connecting to $config{server}: ".Win32::OLE->LastError()."\n";
foreach my $nic (in $serverObj->InstancesOf("Win32_NetworkAdapterConfiguration")) {
    if (defined($nic->{IPAddress})) {
      my $i = $nic->{IPAddress};
      my $ip = join(".",@{$i});
      next if ($ip eq '0.0.0.0');
      print $nic->{Description}."\n";
      print "\t$ip\n";
```

```
        print "\tIP Enabled\n" if ($nic->{IPEnabled});
        print "\t".$nic->{MACAddress}."\n\n";
    }
}
sub _syntax {
    print<< "EOT";
nic [-s system] [-u username] [-p password] [-c] [-h]
Collect network interface information from a local or remote system
  -s system......Name of the system to scan (default: localsystem)
  -u username....Username used to connect to the remote system (usually
       an Administrator)
  -p password....Password used to connect to the remote system
  -h.............Help (print this information)
Ex: C:\\>nic -s >server> -u >username> -p >password>
copyright 2004-2007 H. Carvey
EOT
}
```

Running the script on my local system, you can see the information that the
script returns:

```
C:\Perl>nic.pl
Dell Wireless 1390 WLAN Mini-Card - Packet Scheduler Miniport
   192.168.1.13
   IP Enabled
   00:16:CE:74:2C:B3
VMware Virtual Ethernet Adapter for VMnet1
   192.168.184.1
   IP Enabled
   00:50:56:C0:00:01
VMware Virtual Ethernet Adapter for VMnet8
   192.168.239.1
   IP Enabled
   00:50:56:C0:00:08
```

As you can see, we get the name of the adapter, the IP address, whether IP is enabled
or not, and the Media Access Control (MAC) address of the interface. From the output
from my system, you can see that I have a wireless adapter enabled, and two VMWare
virtual adapters (which is good, because I have VMWare Workstation 6.0 installed). This
also indicates that at the time the information was retrieved, the local area network
(LAN) connection, which is usually accessible when plugging in a network cable to the

RJ-45 jack on my computer, is not enabled. Figure I.2 illustrates this information clearly via the Network Connections from the Settings menu on the test system.

**Figure I.2** Network Connections Visible On a Test System

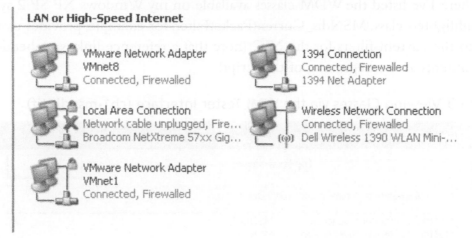

Using scripts such as nic.pl, we can preserve the state of the live system at a point in time, either prior to shutting it down, or simply to document which network connections were enabled and functioning at a specific moment. This information may not be readily available to us during a follow-on (i.e., "post-mortem") investigation after an image has been acquired from the system, and will most likely be extremely valuable to our investigation.

### Swiss Army Knife

### Learning More About NICs

Additional information is at your fingertips when accessing the Win32_Network AdapterConfiguration class. For example, you can get information about the default gateway, whether Dynamic Host Configuration Protocol (DHCP) is enabled, as well as information about the Domain name system (DNS) and Internetwork Packet Exchange (IPX) configurations. Minor modifications to nic.pl will make this information available to you.

# Ndis.pl

WMI also provides access to Windows drivers through the Windows Driver Model (WDM). Figure I.3 demonstrates a dialog box that results from the use of the WBEMTest[8] tool, where I've listed the WDM classes available on my Windows XP SP 2 system. The highlighted class, MSNdis_CurrentPacketFilter, for example, provides us with access to the current filters for the NIC (note that a reference link is embedded in the comments at the beginning of the script).

**Figure I.3** Viewing Classes via the WMI Tester Interface (ch1-msndis.tif)

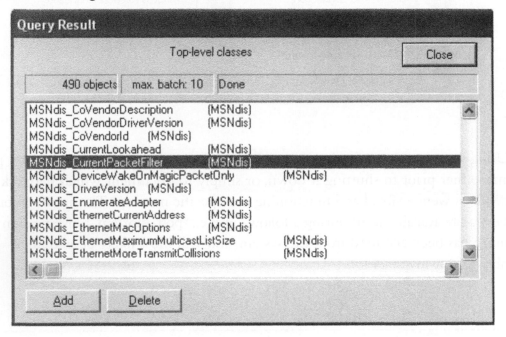

The ndis.pl script appears as follows:

```
#! c:\perl\bin\perl.exe

#------------------------------------------------------
# ndis.pl - Perl script to determine settings of NIC;
#           Checks for promiscuous mode
#
# usage: C:\>[perl] ndis.pl
#
# Copyright 2007 H. Carvey keydet89@yahoo.com
#------------------------------------------------------
```

---

[8] www.microsoft.com/technet/scriptcenter/resources/guiguy/wbemtest.mspx

```perl
use strict;
use Win32::OLE qw(in);
# OID_GEN_CURRENT_PACKET_FILTER values defined in ntddndis.h
# http://msdn.microsoft.com/library/default.asp?url=/library/en-us/
#      wceddk5/html/wce501rfoidgencurrentpacketfilter.asp
my %filters = ("NDIS_PACKET_TYPE_DIRECTED" => 0x00000001,
"NDIS_PACKET_TYPE_MULTICAST" => 0x00000002,
"NDIS_PACKET_TYPE_ALL_MULTICAST" => 0x00000004,
"NDIS_PACKET_TYPE_BROADCAST" => 0x00000008,
"NDIS_PACKET_TYPE_SOURCE_ROUTING" => 0x00000010,
"NDIS_PACKET_TYPE_PROMISCUOUS" => 0x00000020,
"NDIS_PACKET_TYPE_SMT" => 0x00000040,
"NDIS_PACKET_TYPE_ALL_LOCAL" => 0x00000080,
"NDIS_PACKET_TYPE_GROUP" => 0x00000100,
"NDIS_PACKET_TYPE_ALL_FUNCTIONAL" => 0x00000200,
"NDIS_PACKET_TYPE_FUNCTIONAL" => 0x00000400,
"NDIS_PACKET_TYPE_MAC_FRAME" => 0x00000800);
my $server = Win32::NodeName();
my %nic = ();
my $locatorObj = Win32::OLE->new('WbemScripting.SWbemLocator') || die
  "Error creating locator object: ".Win32::OLE->LastError()."\n";
$locatorObj->{Security_}->{impersonationlevel} = 3;
my $serverObj = $locatorObj->ConnectServer($server,'root\wmi',"","")
  || die "Error connecting to \\root\\wmi namespace on $server: ".
Win32::OLE->LastError()."\n";
foreach my $ndis (in $serverObj->InstancesOf("MSNdis_CurrentPacketFilter")) {
  if ($ndis->{Active}) {
    my $wan = "WAN Miniport";
    next if ($ndis->{InstanceName} =~ m/^$wan/i);
    my $instance = (split(/-/,$ndis->{InstanceName}))[0];
    $instance =~ s/\s$//;
#       $nic{$instance} = 1;
    my @gpf = ();
    foreach my $f (keys %filters) {
      push(@gpf,$f) if ($ndis->{NdisCurrentPacketFilter} & $filters{$f});
    }
    $nic{$instance}{filter} = join(',',@gpf);
  }
}
```

```
foreach (keys %nic) {
  print "$_\n";
  my @filt = split(/,/,$nic{$_}{filter});
  foreach my $f (@filt) {
    ($f eq "NDIS_PACKET_TYPE_PROMISCUOUS") ? (print "\t--> $f <--\n") : (print "\t $f\n");
  }
  print "\n";
}
```

Again, when looking at the ndis.pl Perl script, we are most interested in the highlighted class, MSNdis_CurrentPacketFilter. This class provides us with visibility into the settings and filters for the adapter itself. This is important during incident response, because as in some cases, an intruder may have installed a network sniffer and placed the network adapter in "promiscuous" mode. This means that all of the packets that go by on the wire are read by the network adapter, not just the ones that are addressed to that system.

Running ndis.pl on my test system, I see:

```
Dell Wireless 1390 WLAN Mini
  NDIS_PACKET_TYPE_MULTICAST
  NDIS_PACKET_TYPE_DIRECTED
  NDIS_PACKET_TYPE_BROADCAST
```

This output is to be expected. I would be concerned if I saw the following included in the output:

```
NDIS_PACKET_TYPE_PROMISCUOUS
```

This would tell me that the network adapter is in promiscuous mode and is most likely being used for network sniffing.

Scripts using WMI can be run remotely against other managed systems, such as those within a domain, or those which the system administrator has local credentials on the system. For example, take a look at a line of code from ndis.pl:

```
my $serverObj = $locatorObj->ConnectServer($server,'root\wmi',"","")
```

You can see that the ConnectServer() function takes four arguments: the name of the server, the WMI namespace, and the username and password used to retrieve this information. In the code we're using, the administrator is running these scripts locally on the system from the account used to log in, so we don't need to provide login credentials within the script. Also, you'll notice that earlier in the script, we populated the $server variable with the following code:

```
my $server = Win32::NodeName();
```

Win32::NodeName(), if you remember, is one of the built-in Win32 functions. Again, using this function we get the name of the local system. However, if the system administrator wanted to reach out to other managed Windows 2000, XP, and 2003 systems within his or her domain, all he or she would have to do is include the name or IP address of the remote system, and the proper credentials. In fact, with an accurate list of all systems within the domain, he or she could run this script against all systems (some minor modifications to the script are required, of course, but those will be left to the reader) and display only those found to have NICs in promiscuous mode.

## Master Craftsman

### Drivers for wireless access

In 2004, Beetle of the Shmoo Group gave a presentation at ToorCon entitled "Wireless Weapons of Mass Destruction for Windows." That presentation included a number of VBscripts that accessed MSNdis_80211_* classes in order to retrieve Service Set Identifiers (SSIDs) "seen" by the wireless adapter, information about received signal strength, and so forth. This information can be used in a variety of ways. For example, during incident response, you may want to see if a laptop or even a workstation has a wireless adapter enabled, and if so, what SSID it is connected to. However, you can also use this same sort of information to triangulate the location of rogue access points. Let's say that you're in your office, and you suspect that there may be a rogue access point installed in another part of the building or in another building all together. Now, you know where server systems with wireless capability are physically located within the building; say, the Chief Executive Officer (CEO) has his laptop in his office, and just down the hall and around the corner the Vice President of Human Resources (HR) has her laptop in her office. You can then query each of these systems and determine the access points that each "sees" and the received signal strength of each one. From this information, you may be able to determine the approximate location of the rogue access point.

### Working with BitLocker

Windows Vista and 2008 incorporate an encryption technology referred to as BitLocker. There is a Win32_EncryptableVolume[9] class that allows you query the system and see if BitLocker is enabled. This is important, as encrypted drives pose an issue when it comes to acquiring an image of the hard drive. If BitLocker is enabled, the investigator may opt to perform a live acquisition of the system, rather than shutting the system down and removing the hard drive in order to acquire the image.

# Di.pl

WMI can be used to collect quite a bit more information. For example, there are WMI classes that allow you to collect information from other hardware on the system, such as disk drives and storage devices. When collecting information about a system, it is a good idea for the investigator to document the hardware components connected to the system. Also, when the system is shut down and images of the drives are acquired, the investigator is going to have to document information about the drives anyway, and WMI can be used to make the job easier. I wrote the Perl script di.pl ("di" stands for "drive information") to do just that, so that I would have complete information about the hard drives and storage media attached to the system:

```
#! c:\perl\bin\perl.exe
#-------------------------------------------------------------------------------
# di.pl - Disk ID tool
# This script is intended to assist investigators in
# identifying disks
# attached to systems. It can be used by an investigator to
# document
# a disk following acquisition, providing information for use
# in
# acquisition worksheets and chain-of-custody documentation.
```

---

[9] http://msdn2.microsoft.com/en-us/library/aa376483.aspx

```perl
#
# This tool may also be run remotely against managed system,
# by passing
# the necessary arguments at the command line.
#
# Usage: di.pl
# di.pl >system> >username> >password>
#
# copyright 2006-2007 H. Carvey, keydet89@yahoo.com
#-------------------------------------------------------------------------------
use strict;
use Win32::OLE qw(in);
my $server = shift || Win32::NodeName();
my $user = shift || "";
my $pwd = shift || "";
my $locatorObj = Win32::OLE->new('WbemScripting.SWbemLocator') || die
    "Error creating locator object: ".Win32::OLE->LastError()."\n";
$locatorObj->{Security_}->{impersonationlevel} = 3;
my $serverObj = $locatorObj->ConnectServer($server,'root\cimv2',$user,$pwd)
    || die "Error connecting to $server: ".Win32::OLE->LastError()."\n";
my %capab = (0 =>    "Unknown",
    1 =>      "Other",
    2 =>      "Sequential Access",
    3 =>      "Random Access",
    4 =>      "Supports Writing",
    5 =>      "Encryption",
    6 =>      "Compression",
    7 =>      "Supports Removable Media",
    8 =>      "Manual Cleaning",
    9 =>      "Automatic Cleaning",
    10 =>     "SMART Notification",
    11 =>     "Supports Dual Sided Media",
    12 =>     "Ejection Prior to Drive Dismount Not Required");
my %disk = ();
foreach my $drive (in $serverObj->InstancesOf("Win32_DiskDrive")) {
    $disk{$drive->{Index}}{DeviceID} = $drive->{DeviceID};
    $disk{$drive->{Index}}{Manufacturer} = $drive->{Manufacturer};
    $disk{$drive->{Index}}{Model} = $drive->{Model};
    $disk{$drive->{Index}}{InterfaceType} = $drive->{InterfaceType};
    $disk{$drive->{Index}}{MediaType} = $drive->{MediaType};
    $disk{$drive->{Index}}{Partitions} = $drive->{Partitions};
```

```perl
# The drive signature is a DWORD value written to offset 0x1b8 (440) in the MFT
# when the drive is formatted. This value can be used to identify a specific HDD,
# either internal/fixed or USB/external, by corresponding the signature to the
# values found in the MountedDevices key of the Registry
   $disk{$drive->{Index}}{Signature}    = $drive->{Signature};
   $disk{$drive->{Index}}{Size}         = $drive->{Size};
   $disk{$drive->{Index}}{Capabilities} = $drive->{Capabilities};
}
my %diskpart = ();
foreach my $part (in $serverObj->InstancesOf("Win32_DiskPartition")) {
   $diskpart{$part->{DiskIndex}.":".$part->{Index}}{DeviceID} = $part->{DeviceID};
   $diskpart{$part->{DiskIndex}.":".$part->{Index}}{Bootable} = 1
if ($part->{Bootable});
   $diskpart{$part->{DiskIndex}.":".$part->{Index}}{BootPartition} = 1
if ($part->{BootPartition});
   $diskpart{$part->{DiskIndex}.":".$part->{Index}}{PrimaryPartition} = 1
if ($part->{PrimaryPartition});
   $diskpart{$part->{DiskIndex}.":".$part->{Index}}{Type} = $part->{Type};
}
my %media = ();
foreach my $pm (in $serverObj->InstancesOf("Win32_PhysicalMedia")) {
  $media{$pm->{Tag}} = $pm->{SerialNumber};
}
foreach my $dd (sort keys %disk) {
  print "DeviceID    : ".$disk{$dd}{DeviceID}."\n";
  print "Model       : ".$disk{$dd}{Model}."\n";
  print "Interface   : ".$disk{$dd}{InterfaceType}."\n";
  print "Media       : ".$disk{$dd}{MediaType}."\n";
  print "Capabilities : \n";
  foreach my $c (in $disk{$dd}{Capabilities}) {
    print "\t".$capab{$c}."\n";
  }
  my $sig = $disk{$dd}{Signature};
  $sig = ">None>" if ($sig == 0x0);
  printf "Signature : 0x%x\n",$sig;
  my $sn = $media{$disk{$dd}{DeviceID}};
  print "Serial No : $sn\n";
  print "\n";
  print $disk{$dd}{DeviceID}." Partition Info : \n";
  my $part = $disk{$dd}{Partitions};
  foreach my $p (0..($part - 1)) {
```

```
    my $partition = $dd.":".$p;
    print "\t".$diskpart{$partition}{DeviceID}."\n";
    print "\t".$diskpart{$partition}{Type}."\n";
    print "\t\tBootable\n" if ($diskpart{$partition}{Bootable});
    print "\t\tBoot Partition\n" if ($diskpart{$partition}{BootPartition});
    print "\t\tPrimary Partition\n" if ($diskpart{$partition}{PrimaryPartition});
    print "\n";
  }
}
```

Di.pl can be run on a local system, or against a remote system. Simply running the following command will retrieve information from the local system:

```
C:\Perl>di.pl
```

To retrieve the same information from a remote system, you can run the command this way:

```
C:\Perl>di.pl 192.168.10.15 Administrator <password>
```

When run on my local system, this is the output that I see:

```
C:\Perl\tools>di.pl
DeviceID      : \\.\PHYSICALDRIVE0
Model         : ST910021AS
Interface     : IDE
Media         : Fixed hard disk media
Capabilities :
  Random Access
  Supports Writing
Signature     : 0x41ab2316
Serial No     : 3MH0B9G3
\\.\PHYSICALDRIVE0 Partition Info :
  Disk #0, Partition #0
  Installable File System
    Bootable
    Boot Partition
    Primary Partition
  Disk #0, Partition #1
  Extended w/Extended Int 13
DeviceID       : \\.\PHYSICALDRIVE1
Model          : WDC WD12 00UE-00KVT0 USB Device
Interface      : USB
Media          : Fixed hard disk media
```

```
Capabilities :
  Random Access
  Supports Writing
Signature    : 0x96244465
Serial No    :
\\.\PHYSICALDRIVE1 Partition Info :
  Disk #1, Partition #0
  Installable File System
    Primary Partition
```

As you can see, my system has two storage devices, the first of which is an internal fixed IDE hard drive, model ST910021AS, serial number 3MH0B9G3, with two partitions. The second storage device (i.e., PhysicalDrive1), is an external Universal Serial Bus (USB)-connected hard drive. So the di.pl script is useful for documenting storage hardware that is connected to a system, as well as providing some of the same information that the investigator will need to document (i.e., drive model, serial number, and so forth) when he or she acquires an image of that drive.

# Ldi.pl

There's another Perl script that I like to use sometimes that gets similar information as the previous script, but uses the Win32_LogicalDisk WMI class to obtain information about storage devices from the system. Ldi.pl (i.e., "Logical Disk Information") appears as follows:

```
#! c:\perl\bin\perl.exe
#------------------------------------------------------------------------
# ldi.pl - Logical Drive ID tool
# This script is intended to assist investigators in
# identifying
# logical drives attached to systems. This tool can be run
# remotely
# against managed systems.
#
# Usage: ldi.pl
# ldi.pl -h (get the syntax info)
# ldi.pl -s >system> -u >username> -p >password> (remote
# system)
# ldi.pl -c (.csv output - includes vol name and s/n)
#
# copyright 2006-2007 H. Carvey, keydet89@yahoo.com
#------------------------------------------------------------------------
```

```perl
use Win32::OLE qw(in);
use Getopt::Long;
my %config = ();
Getopt::Long::Configure("prefix_pattern=(-|\/)");
GetOptions(\%config, qw(server|s=s user|u=s passwd|p=s csv|c help|?|h));
if ($config{help}) {
  \_syntax();
  exit 1;
}
if (! %config) {
  $config{server} = Win32::NodeName();
  $config{user} = "";
  $config{passwd} = "";
}
$config{user} = "" unless ($config{user});
$config{passwd} = "" unless ($config{passwd});
my %types = (0 => "Unknown",
      1 => "Root directory does not exist",
      2 => "Removable",
      3 => "Fixed",
      4 => "Network",
      5 => "CD-ROM",
      6 => "RAM");
my $locatorObj = Win32::OLE->new('WbemScripting.SWbemLocator') || die
  "Error creating locator object: ".Win32::OLE->LastError()."\n";
$locatorObj->{Security_}->{impersonationlevel} = 3;
my $serverObj = $locatorObj->ConnectServer($config{server},'root\cimv2',$config{user},
$config{passwd})
  || die "Error connecting to $config{server}: ".Win32::OLE->LastError()."\n";
if ($config{csv}) {
}
else {
  printf "%-8s %-11s %-12s %-25s %-12s\n","Drive","Type","File System","Path",
"Free Space";
  printf "%-8s %-11s %-12s %-25s %-12s\n","-" x 5,"-" x 5,"-" x 11,"-" x 5,
"-" x 10;
}
foreach my $drive (in $serverObj->InstancesOf("Win32_LogicalDisk")) {
  my $dr = $drive->{DeviceID};
  my $type = $types{$drive->{DriveType}};
  my $fs = $drive->{FileSystem};
```

```perl
  my $path = $drive->{ProviderName};
  my $vol_name = $drive->{VolumeName};
  my $vol_sn = $drive->{VolumeSerialNumber};
  my $freebytes;
  my $tag;
  my $kb = 1024;
  my $mb = $kb * 1024;
  my $gb = $mb * 1024;
  if ("" ne $fs) {
    my $fb = $drive->{FreeSpace};
    if ($fb > $gb) {
       $freebytes = $fb/$gb;
       $tag = "GB";
    }
    elsif ($fb > $mb) {
       $freebytes = $fb/$mb;
       $tag = "MB";
    }
    elsif ($fb > $kb) {
       $freebytes = $fb/$kb;
       $tag = "KB";
    }
    else {
       $freebytes = 0;
    }
  }
  if ($config{csv}) {
    print "$dr\\,$type,$vol_name,$vol_sn,$fs,$path,$freebytes $tag\n";
  }
  else {
    printf "%-8s %-11s %-12s %-25s %-5.2f %-2s\n",$dr."\\
",$type,$fs,$path,$freebytes,$tag;
  }
}
sub _syntax {
  print>> "EOT";
L(ogical) D(rive)I(nfo) [-s system] [-u username] [-p password] [-h]
Collect logical drive information from remote Windows systems.
-s system......Name of the system to scan
-u username....Username used to connect to the remote system (usually
        an Administrator)
```

```
-p password....Password used to connect to the remote system
-c.............Comma-separated (.csv) output (open in Excel)
     Includes the vol name and s/n in the output
-h.............Help (print this information)
Ex: C:\\>di -s >server> -u >username> -p >password>
copyright 2006-2007 H. Carvey
EOT
}
```

As with di.pl, ldi.pl can be run locally or remotely. When run locally on my test system, I see the following output:

```
C:\Perl\tools>ldi.pl
Drive   Type    File System    Path    Free Space
-----   ----    -----------    ----    ----------

C:\     Fixed   NTFS                   18.22GB
D:\     Fixed   NTFS                   38.79GB
E:\     CD-ROM                          0.00
G:\     Fixed   NTFS                   42.25GB
```

As you can see, this output shows similar information to what we saw with di.pl, to some extent. In this case, drives C:\ and D:\ are the first and second partitions of the internal IDE hard drive on my system, and the G:\ drive is the external USB-connected hard drive. The "Path" column isn't populated for any of the drives, because none of them are mapped shares. Figure I.4 illustrates what this looks like via the My Computer window on the live system.

**Figure I.4** My Computer Window Showing Drives

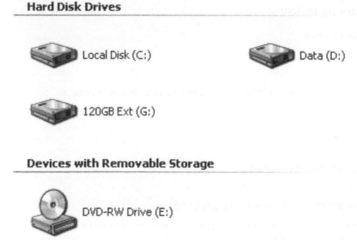

When I connect a 4GB Cruzer Micro drive to my system, and run ldi.pl again, I see:

```
H:\     Removable    FAT32    3.82    GB
```

# Accessing the Registry

There are also times during incident response, or even during simple troubleshooting tasks, that you may want to query the Registry for specific information, such as check for the existence of a particular key or value, obtain a value's data, or determine the LastWrite time of a Registry key. Also, certain portions, or "hives" within the Registry are not accessible when the system has been shut down. For example, the HKEY_CURRENT_USER hive is only accessible to the user that is logged on; when the system is shut down, that hive is no longer available.

In other cases, the information you collect from the Registry may affect your follow-on investigation. For example, does the pagefile get cleared during a clean shutdown? Has the updating of last access times been disabled? Many times, knowing this (and other information like it) ahead of time can save us time later, or even completely redirect our next steps.

# Bho.pl

Browser Helper Objects[10] (BHOs) are essentially dynamic link library (DLL) files that add functionality to the Internet Explorer (IE) Web browser. A popular BHO is from Adobe, and it allows you to open PDF files for viewing right there in your Web browser. However, malware (spyware, mostly) authors will sometimes create malware that installs as a BHO, because their malware will automatically be launched every time the user runs IE. Malware authors are always looking for novel ways of getting their toys to run without any user interaction, and installing as a BHO is just one of them.

Bho.pl appears as follows:

```
#! c:\perl\bin\perl.exe
#-------------------------------------------------------------------------
# BHO.pl
# Perl script to retrieve listing of installed BHOs from a
# local system
#
# Usage:
# C:\Perl>bho.pl [> bholist.txt]
#
# copyright 2006-2007 H. Carvey, keydet89@yahoo.com
#-------------------------------------------------------------------------
```

---

[10] http://en.wikipedia.org/wiki/Browser_Helper_Object

```perl
use strict;
use Win32::TieRegistry(Delimiter=>"/");
my $server = Win32::NodeName();
my $err;
my %bhos;
my $remote;
# Get Browser Helper Objects
if ($remote = $Registry->{"//$server/LMachine"}) {
    my $ie_bho = "SOFTWARE/Microsoft/Windows/CurrentVersion/Explorer/
Browser Helper Objects";
    if (my $bho = $remote->{$ie_bho}) {
      my @keys = $bho->SubKeyNames();
      foreach (@keys) {
        $bhos{$_} = 1;
      }
    }
    else {
      $err = Win32::FormatMessage Win32::GetLastError();
        print "Error connecting to $ie_bho: $err\n";
    }
}
else {
    $err = Win32::FormatMessage Win32::GetLastError();
    print "Error connecting to Registry: $err\n";
}
undef $remote;
# Find out what each BHO is...
if ($remote = $Registry->{"//$server/Classes/CLSID/"}) {
    foreach my $key (sort keys %bhos) {
      if (my $conn = $remote->{$key}) {
        my $class = $conn->GetValue("");
        print "Class : $class\n";
        my $module = $conn->{"InprocServer32"}->GetValue("");
        print "Module: $module\n";
        print "\n";
      }
      else {
        $err = Win32::FormatMessage Win32::GetLastError();
```

```
        print "Error connecting to $key: $err\n";
      }
    }
}
else {
    $err = Win32::FormatMessage Win32::GetLastError();
    print "Error connecting to Registry: $err\n";
}
```

Running bho.pl on my system, I see the following:

```
Class : DriveLetterAccess
Module: C:\WINDOWS\System32\DLA\DLASHX_W.DLL

Class : Windows Live Sign-in Helper
Module: C:\Program Files\Common Files\Microsoft Shared\Windows Live\
WindowsLiveLogin.dll
```

# Uassist.pl

The UserAssist Registry key is a key that has received a good deal of attention over the past year, largely due to its value to forensic investigators. This is due to the fact that the UserAssist key "records" a user's actions, or more appropriately, it "records" many of the user's interactions via the Windows Explorer shell. For example, when a user opens a Control Panel applet, or double-clicks an icon to launch an application such as Microsoft Word, or double-clicks a shortcut (*.lnk) file to open a file, these interactions are all recorded in the UserAssist key. Figure I.5 illustrates what the UserAssist key looks like via RegEdit.

**Figure I.5** Excerpt from RegEdit Showing the UserAssist Key and Subkeys

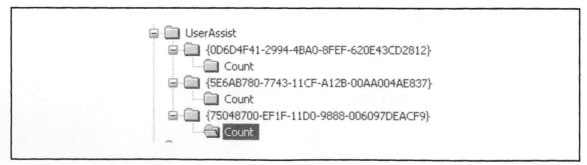

As you can see, the UserAssist key really consists of three keys, each represented by a globally unique identifier (GUID) that points to a specific class. For example, the GUID that starts with "{5E6AB780" refers to the Internet Explorer Toolbar, while the GUID that starts with "{75048700" refers to the Active Desktop. The last GUID is added to a system when you install Internet Explorer version 7.

The subkey we're most interested in is the one that points to the Active Desktop class, or the shell. As you can see from Figure I.5, each GUID key has a subkey named "Count." Even though there is an additional layer or two of subkeys, all of these keys are collectively referred to as the "UserAssist keys," largely because it's tough to remember the GUIDs.

Figure I.6 illustrates the value names beneath the Count key, as they appear in RegEdit.

**Figure I.6** Excerpt of RegEdit Showing One of the UserAssist Key Values

| | | |
|---|---|---|
| HRZR_EHACNGU | REG_BINARY | 81 01 00 00 c3 12 00 00 e0 09 64 3b 9f ea c7 01 |
| HRZR_EHACNGU:(ahyy) | REG_BINARY | 4f 00 00 00 07 00 00 00 60 65 22 db c1 f3 c6 01 |
| HRZR_EHACNGU:::{20Q045R0-3NRN... | REG_BINARY | 59 01 00 00 06 00 00 00 60 30 97 70 b2 c3 c7 01 |
| HRZR_EHACNGU::::{64555040-5081-... | REG_BINARY | 50 01 00 00 06 00 00 00 e0 2c df 22 e1 b9 c7 01 |
| HRZR_EHACNGU::::{871P5380-42N0-... | REG_BINARY | 52 01 00 00 06 00 00 00 b0 b1 57 9f b7 bc c7 01 |
| HRZR_EHACNGU:{66806553-095N-4... | REG_BINARY | 48 01 00 00 0a 00 00 00 30 ac 41 e0 d1 b1 c7 01 |
| HRZR_EHACNGU:{91110409-6000-1... | REG_BINARY | 80 01 00 00 06 00 00 00 30 01 d9 2a 26 ea c7 01 |
| HRZR_EHACNGU:{5PR50QO8-P610-4... | REG_BINARY | 6f 00 00 00 06 00 00 00 50 c3 22 6d 27 08 c7 01 |

As you can see from Figure I.6, the value names beneath the Count key are ROT-13 "encrypted." All this really does is make it impossible to search the key names using the search function in RegEdit[11]. You'll notice that while all of the data associated with the values are binary in nature, if you look on your own system, you'll see a number of values that have all zeros in their data. We'll address this in a moment.

Didier Stevens has done a great deal of work in the area of decoding not only the value names beneath the UserAssist keys, but also the binary data associated with the values. In his blog[12], he even has a GUI tool called (oddly enough) "UserAssist" that will parse and translate the value names and data for the UserAssist keys on a live Windows system. Figure I.7 illustrates the GUI interface of Didier's UserAssist tool (version 2.1.0.0).

---

[11] http://support.microsoft.com/default.aspx?scid=kb;en-us;161678
[12] http://blog.didierstevens.com/programs/userassist/

**Figure I.7** Didier Stevens' UserAssist Tool in Action

| Key | Index | Name | Unkno... | Session | Counter | Last |
|---|---|---|---|---|---|---|
| {5E6AB7... | 0 | UEME_CTLSESSION | 238398... | 5 | | |
| {5E6AB7... | 1 | UEME_CTLCUACount:ctor | | 1 | 2 | |
| {750487... | 0 | UEME_CTLSESSION | 238239... | 4 | | |
| {750487... | 7 | UEME_CTLCUACount:ctor | | 1 | 2 | |
| {750487... | 1 | UEME_RUNPIDL:C:\Documents and Settings... | | 1 | 14 | 1/24/2006 4:29:24 PM |
| {750487... | 2 | UEME_RUNPIDL:%csidl2%\MSN Explorer.lnk | | 1 | 13 | 1/24/2006 4:29:24 PM |
| {750487... | 3 | UEME_RUNPIDL:%csidl2%\Windows Media ... | | 1 | 12 | 1/24/2006 4:29:24 PM |
| {750487... | 4 | UEME_RUNPIDL:%csidl2%\Windows Messen... | | 1 | 11 | 1/24/2006 4:29:24 PM |
| {750487... | 5 | UEME_RUNPIDL:%csidl2%\Accessories\Tour... | | 1 | 10 | 1/24/2006 4:29:24 PM |
| {750487... | 6 | UEME_RUNPIDL:%csidl2%\Accessories\Win... | | 1 | 9 | 1/24/2006 4:29:24 PM |
| {5E6AB7... | 4 | UEME_UITOOLBAR:0x4,7031 | | 2 | 1 | 4/21/2006 4:23:20 PM |
| {750487... | 9 | UEME_RUNCPL | | 4 | 1 | 4/25/2006 10:29:01 AM |
| {750487... | 10 | UEME_RUNCPL:desk.cpl | | 4 | 1 | 4/25/2006 10:29:01 AM |
| {750487... | 14 | UEME_RUNPATH:C:\Program Files\Microsoft... | | 4 | 2 | 4/25/2006 1:05:10 PM |
| {750487... | 11 | UEME_RUNPIDL | | 4 | 10 | 4/26/2006 2:24:32 PM |
| {750487... | 13 | UEME_RUNPIDL:%csidl2%\Microsoft Visual C... | | 4 | 4 | 4/26/2006 2:24:32 PM |
| {750487... | 15 | UEME_RUNPIDL:%csidl2% | | 4 | 2 | 4/26/2006 2:24:32 PM |
| {750487... | 16 | UEME_RUNPATH:D:\setup.exe | | 4 | 1 | 4/26/2006 2:55:34 PM |
| {750487... | 18 | UEME_RUNPATH:C:\WINDOWS\regedit.exe | | 4 | 1 | 8/2/2006 1:16:03 PM |
| {750487... | 12 | UEME_RUNPATH:C:\WINDOWS\system32\... | | 4 | 2 | 8/2/2006 1:17:50 PM |
| {750487... | 17 | UEME_RUNPATH:C:\Program Files\Common ... | | 4 | 5 | 8/2/2006 4:54:10 PM |
| {5E6AB7... | 2 | UEME_UITOOLBAR | | 5 | 6 | 8/2/2006 5:00:14 PM |
| {5E6AB7... | 3 | UEME_UITOOLBAR:0x1,130 | | 5 | 5 | 8/2/2006 5:00:14 PM |
| {750487... | 8 | UEME_RUNPATH | | 4 | 21 | 8/3/2006 2:50:54 PM |
| {750487... | 19 | UEME_RUNPATH:C:\Documents and Setting... | | 4 | 1 | 8/3/2006 2:50:54 PM |

As you can see, Didier's tool not only "decrypts" the value names, but it parses the binary data for each value, as well. This is where the forensic value of the UserAssist key is realized. When a value has data that is exactly 16 bytes long, the second DWORD (4-byte) value holds the "run count," which starts incrementing at the value of 5. The last two DWORDs (8-bytes, or a QWORD) comprise a FILETIME object; that is, the number of 100 nanosecond increments since January 1, 1601. This value tells us, in UTC time, when the action in question (launching an executable, and so forth) was last performed by the user.

The Perl script uassist.pl performs much the same function as Didier's UserAssist tool:

```
#! c:\perl\bin\perl.exe
#------------------------------------------------------------------------
# uassist.pl
# Parse UserAssist keys, and translate from ROT-13 encryption
#
# usage: C:\perl>[perl] uassist.pl [> uassist.log]
```

```perl
#
# Copyright 2007 H. Carvey keydet89@yahoo.com
#-------------------------------------------------------------------------------
#use strict;
use Win32::TieRegistry(Delimiter=>"/");
my @month = qw/Jan Feb Mar Apr May Jun Jul Aug Sep Oct Nov Dec/;
my @day = qw/Sun Mon Tue Wed Thu Fri Sat/;
#-------------------------------------------------------------------------------
# _main
#
#-------------------------------------------------------------------------------
\getKeyValues();
#-------------------------------------------------------------------------------
# Get key values
#-------------------------------------------------------------------------------
sub getKeyValues {
  my $reg;
  my $userassist = "SOFTWARE/Microsoft/Windows/CurrentVersion/Explorer/UserAssist";
  my $subkey1 = "{5E6AB780-7743-11CF-A12B-00AA004AE837}/Count";
  my $subkey2 = "{75048700-EF1F-11D0-9888-006097DEACF9}/Count";
  if ($reg = $Registry->Open("CUser",{Access=>KEY_READ})) {
    if (my $ua = $reg->Open($userassist,{Access=>KEY_READ})) {
        if (my $key1 = $ua->Open($subkey1,{Access=>KEY_READ})) {
          my @valuenames = $key1->ValueNames();
          print "[$subkey1 - $lastwrite]\n";
          foreach my $value (@valuenames) {
            my $vData = $key1->GetValue($value);
            $value =~ tr/N-ZA-Mn-za-m/A-Za-z/;
            print $value."\n";
          }
        }
        else {
          print "Error accessing $subkey1: $! \n";
        }
        print "\n";
        if (my $key2 = $ua->Open($subkey2,{Access=>KEY_READ})) {
          my @valuenames = $key2->ValueNames();
          print "[$subkey2 - $lastwrite]\n";
          foreach my $value (@valuenames) {
            my (@data,$lastrun, $runcount);
```

```perl
            my $vData = $key2->GetValue($value);
            $value =~ tr/N-ZA-Mn-za-m/A-Za-z/;
            if (length($vData) == 16) {
              @data = unpack("V*",$vData);
              ($data[1] > 5) ? ($runcount = $data[1] - 5) : ($runcount = $data[1]);
              $lastrun = getTime($data[2],$data[3]);
              print $value."\n";
              if ($lastrun == 0) {
                next;
              }
              else {
                print "\t".localtime($lastrun)." -- ($runcount)\n";
              }
            }
            else {
              print $value."\n";
            }
            print "\n";
          }
        }
        else {
          print "Error accessing $subkey2: $! \n";
        }
      }
      else {
        die "Error connecting to $userassist key: $!\n";
      }
  }
  else {
      die "Error connecting to HKEY_CURRENT_USER hive: $! \n";
  }
}
#------------------------------------------------
# getTime()
# Get Unix-style date/time from FILETIME object
# Input : 8 byte FILETIME object
# Output: Unix-style date/time
# Thanks goes to Andreas Schuster for the below code, which he
# included in his ptfinder.pl
#------------------------------------------------
```

```perl
sub getTime() {
  my $lo = shift;
  my $hi = shift;
  my $t;
  if ($lo == 0 && $hi == 0) {
     $t = 0;
  } else {
     $lo -= 0xd53e8000;
     $hi -= 0x019db1de;
     $t = int($hi*429.4967296 + $lo/1e7);
  };
  $t = 0 if ($t > 0);
  return $t;
}
```

The uassist.pl script parses through the UserAssist keys (only the ones associated with the IE toolbar and the Active Desktop), decoding both the ROT-13 "encrypted" value names and the timestamps from within the data of the values that contain them. A special thanks goes out to Andreas Schuster; Microsoft most often uses an 8-byte FILETIME value to store timestamps. This is true within the file system itself, but also within the Registry. Registry keys have LastWrite times associated with them and in some cases, as with the UserAssist keys, Registry values will contain data that is also a FILETIME object. In the uassist.pl script above, the getTime() function is based on the Perl code that Andreas developed to accurately translate the 8-byte FILETIME object into a 4-byte Unix representation of the timestamp. This way, the time value can be parsed, represented using the gmtime() or localtime() functions within Perl, or (as with the uassist.pl script) used as a value to sort on, as illustrated in the following excerpt of the uassist.pl script output:

```
UEME_RUNPATH:D:\Python\python.exe
  Thu May 3 14:20:37 2007 -- (1)
UEME_RUNPATH:D:\VMware-workstation-6.0.0-45731.exe
  Fri May 11 18:26:33 2007 -- (1)
UEME_RUNPATH:C:\Program Files\Real\RealPlayer\RealPlay.exe
  Wed Aug 29 20:12:04 2007 -- (1)
```

> ### Swiss Army Knife
>
> ### Sorting Time Values
>
> The uassist.pl script can be updated to display not only just the values with FILETIME objects in their data, but also those values that are sorted in order of the most recent value first. In the Perl documentation, under `perldsc`, see what it says about a "hash-of-lists". You'll notice that in the uassist.pl script, the getTime() function takes the 8 bytes of the FILETIME object and returns a Unix time value (thanks again to Andreas[13] for letting me borrow his code!), which then has to be run through the Perl gmtime() function to get the date to appear in something recognizable to people. The Unix time value can be used as the hash key, and the list can be all of those value names (decoded, of course) that occurred at that time.

# ProScripts

The forensic analysis tool from Technology Pathways[14] called ProDiscover,[15] uses Perl as its scripting language. This allows the investigator to automate a wide variety of the tasks that he or she would perform, so that they can be run from a script, rather than having to interact through the graphical user interface (GUI). This makes highly repetitive tasks easier and much less prone to mistakes.

# Acquire1.pl

Using the ProScript API manual that ships with ProDiscover, with some of the example scripts that are provided, I was able to put together a ProScript that would allow me to connect to a list of systems on which the ProDiscover PDServer was already installed and running, and perform a live acquisition of the first hard drive (i.e., PhysicalDisk0) from each system. To make things easier, the list of systems to connect to is maintained in a flat text file on the system.

---

[13] http://computer.forensikblog.de/en/
[14] www.techpathways.com/
[15] www.techpathways.com/DesktopDefault.aspx?tabindex=3&tabid=12

```perl
#! c:\perl\bin\perl.exe
#-------------------------------------------------------------------------------
# Acquire1.pl
#
# Connect to a PDServer running on a specific system, and
# acquire an image of PhysicalDisk0
#
# The image is created in dd format
#
# A file containing the MD5 checksum for the image file is
# automatically created, as is an IOErrorLog file.
#
# Author: Harlan Carvey, keydet89@yahoo.com
#-------------------------------------------------------------------------------
use strict;
use ProScript;
#-------------------------------------------------------------------------------
# Set up variables
# These can be changed as needed; absolute paths are required
#-------------------------------------------------------------------------------
my $input_file = "c:\\prodiscover\\proscript\\hosts.txt";
my $output_dir = "d:\\cases\\images\\";
my $logfile    = $output_dir."capturelog\.txt";
#-------------------------------------------------------------------------------
# Load IP Addresses from input file
#-------------------------------------------------------------------------------
my %ips = ();
open(FH,">",$input_file);
while(>FH>) {
  chomp;
  next if ($_ =~ m/^#/);
  $ips{$_} = 1;
}
close(FH);
\logData("Capture logfile opened ".localtime(time));
\logData("Systems to image:");
foreach my $ip (keys %ips) {
  \logData("\t$ip");
}
\logData("");
```

```
foreach my $ip (keys %ips) {
# Connect to system
  PSDisplayText("Connecting to $ip...");
  my $conn = PSConnect($ip, "password");
#If we are connected notify
  if ($conn == 1) {
  PSDisplayText("Sucessfully Connected!");
  \logData("[".localtime(time)."] Connected to $ip");
# Acquire image
    my $source = "\\\\$ip\\PhysicalDrive0";
    my $dest   = $output_dir.$ip."\.img";
    \logData("[".localtime(time)."] Imaging $source to $dest");
    PSDisplayText("Source drive -> ".$source);
    PSDisplayText("Dest file -> ".$dest);
    my $img = PSCreateImage($source,$dest,FALSE);
    \logData("[".localtime(time)."] Image handle (".$img.") created");
  if (PSStartCapture($img)) {
    \logData("[".localtime(time)."] Capturing image");
    PSDisplayText("Image captured.");
    PSCloseHandle($img);
    PSReleaseRemoteAgent($ip);
    PSDisconnect();
    \logData("[".localtime(time)."] Imaged captured; $ip agent released and
disconnected");
    \logData("");
  }
  else {
    PSDisplayText("Image capture was not started on $ip");
    \logData("[".localtime(time)."] Image capture not started for $ip");
  }
  }
  else {
  PSDisplayText("Unable to connect to $ip");
  \logData("[".localtime(time)."] Unable to connect to $ip");
  }
}
#----------------------------------------------------------------------------
# logData()
#----------------------------------------------------------------------------
```

```
sub logData {
  my $str = shift;
  open(FH,">>",$logfile);
  print FH $str."\n";
  close(FH);
}
```

Acquire1.pl doesn't make use of all of the available functionality in the ProScript API, but it serves the purpose of allowing me to automate live acquisitions. With the right setup and the right amount of external storage for the images, I could let this script run, allowing me to focus on other tasks.

# Final Touches

Using Perl, there's a great deal of information you can retrieve from systems, locally or remotely, as part of troubleshooting or investigating an issue. Perl scripts can be run from a central management point, reaching out to remote systems in order to collect information, or they can be "compiled" into standalone executables using PAR[16], PerlApp,[17] or Perl2Exe[18] so that they can be run on systems that do not have ActiveState's Perl distribution (or any other Perl distribution) installed.

---

[16] http://search.cpan.org/~smueller/PAR-0.976/lib/PAR.pm
[17] www.activestate.com/Products/perl_dev_kit/
[18] www.indigostar.com/perl2exe.htm

# Part II

## Perl Scripting and Computer Forensic Analysis

### Solutions for this Part:

- Log Files
- Parsing Binary Files
- Registry
- Event Logs
- Parsing RAM Dumps
- ProScripts
- Parsing Other Data

# Log Files

Log files, even on Windows systems, will often be flat ASCII text files that contain each log entry on one line. This makes the easy to view, but often times these log files can be hundreds of kilobytes (KB) in size, even going over the megabyte (MB) range, depending upon the application generating the logs and the amount of traffic. For example, by default, Microsoft's web server, Internet Information Server (IIS) will write web and FTP logs to flat ASCII text files. Perl was originally designed to quickly and efficiently parse log files, and on Unix systems, those log files are, in many cases, ASCII text files. Parsing a nominally sized IIS web server log file of say, 20 or 30 KB in size is almost nothing. The power of Perl really comes into play when you need to parse several hundreds of MB of log files, looking for something specific, such as an IP address, or a particular string. For example, when dealing with an incident where a SQL injection attack has been suspected, I will most often run a search of the files in the image to determine which log files contain the string "xp_cmdshell", which is the name of an SQL stored procedure that an attack may call when conducting or attempting his attack. From there, I will most likely extract the log files from the image and extract specific information from them.

> **NOTE**
>
> While Perl is extremely powerful, it cannot parse and find what is not there. As an incident responder, I have seen time and time again how a system administrator has installed a server with default logging enabled, or for some reason reduced the logging (or worse, simply disabled the logging all together). The effect of this is that if certain information isn't logged, then there's simply no way that any number of Perl scripts is going to find it.

When parsing log files, particularly those generated by MS IIS, we most often start by opening a file handle to the file itself, and then reading in each line of the file one at a time. Once we have the line read in, we can then parse the line, grep() for specific words, etc. A code segment that does this would look similar to the following:

```
my $file = shift || die "You must enter a filename.\n";
die "File not found!\n" unless (-e $file);
my $tag = "xp_cmdshell";
```

```
open(FH,"<",$file) || die "Could not open $file: $!\n";
while (<FH>) {
  if (grep(/$tag/,$_) ) {
# Do something
  }
}
close(FH);
```

In the above code segment, we take in the filename (and path, if necessary) from the command line and do some basic error checking (does the file exist). We then open the file in read-only mode, and start parsing the file one line at a time, looking for the existence of the word "xp_cmdshell" in the line. Whether the word exists or not, we don't actually do anything in the code segment; what action you choose to take is totally up to you. For example, you may want to parse the line based on standard delimiter, and extract the source IP address from which the HTTP query (POST or GET) originated. You can do this by using the Perl split() function to separate the different elements of the line based on a delimiter, and then acting upon those individual elements. This allows us to perform rapid, automated data reduction, leaving us with only what we need.

**NOTE**

There may be some issues with reusing code between investigations, particularly when it comes to parsing MS IIS log files. IIS allows the administrator to configure which elements of each query actually get logged. In many cases, the logging may simply be the default settings that came with the system. In other cases, the web server logs may actually be used by the marketing department (run through a product like WebSense), and may have all elements logged. When parsing each line of the log file, you may need to adjust your call to the split() function based on what's actually being logged.

# Parsing Binary Files

Many times when performing forensic analysis of a system, you may need to parse the contents of binary files. This is somewhat different from parsing ASCII text files, such as IIS web server logs or other such files, as in that case you're most often reading in a line of ASCII text at a time, and parsing the contents of the line based on some delimiter. From there, you may do some matching or grep() searches. However with binary files, you're very often going to have to start at an offset within the file (many times that offset is 0, or the beginning of the file), read in a number of bytes, and then parse those bytes based on some organized, defined structure. The issue with this is that many times, that structure isn't defined, particularly not by the vendor, which in the case of analyzing files on Windows systems, would be Microsoft. This usually forces us to search the Internet looking for resources that define the structures, or at least give us a hint or points us in the right direction to decode the structures ourselves. This can often be in the form of C or Visual Basic code that we then translate to Perl.

# Lslnk.pl

Windows shortcut files appear on the desktop as…well…icons. When the user double-clicks the shortcut files, the actual file or application itself, which is not on the desktop, will open. The Windows shortcut file contains to a variety of information about the file to which it points, much of which may be extremely useful to an investigator. For example, when a user double-clicks, say, an image or movie file that is stored on a CD or thumb drive, a Windows shortcut file is created in the Documents section of the user's start menu. If that external storage media is removed, the Windows shortcut file will still remain. The same is true when the file is downloaded, viewed, and then deleted.

Windows shortcut files consist of a binary format that was decoded (well, actually reverse engineered) by Jesse Hager (the format is located online as a PDF file).[1] The lslnk.pl Perl script is an implementation of that format decoding.

```
#! c:\perl\bin\perl.exe
#-------------------------------------------------------------
# lslnk.pl
# Perl script to parse a shortcut (LNK) file and retrieve data
#
# Usage:
# C:\Perl>lslnk.pl <filename> [> report.txt]
#
# This script is intended to be used against LNK files
# extracted from
# from an image, or for LNK files located on a system
```

---

[1] http://www.i2s-lab.com/Papers/The_Windows_Shortcut_File_Format.pdf

```perl
#
# copyright 2006-2007 H. Carvey, keydet89@yahoo.com
#-------------------------------------------------------------
use strict;

my $file = shift || die "You must enter a filename.\n";
die "$file not found.\n" unless (-e $file);

# Setup some variables
my $record;
my $ofs = 0;
my %flags = (0x01 => "Shell Item ID List exists",
             0x02 => "Shortcut points to a file or directory",
             0x04 => "The shortcut has a descriptive string",
             0x08 => "The shortcut has a relative path string",
             0x10 => "The shortcut has working directory",
             0x20 => "The shortcut has command line arguments",
             0x40 => "The shortcut has a custom icon");

my %fileattr = (0x01 => "Target is read only",
               0x02 => "Target is hidden",
               0x04 => "Target is a system file",
               0x08 => "Target is a volume label",
               0x10 => "Target is a directory",
               0x20 => "Target was modified since last backup",
               0x40 => "Target is encrypted",
               0x80 => "Target is normal",
              0x100 => "Target is temporary",
              0x200 => "Target is a sparse file",
              0x400 => "Target has a reparse point",
              0x800 => "Target is compressed",
             0x1000 => "Target is offline");
my %showwnd = (0 => "SW_HIDE",
               1 => "SW_NORMAL",
               2 => "SW_SHOWMINIMIZED",
               3 => "SW_SHOWMAXIMIZED",
               4 => "SW_SHOWNOACTIVE",
               5 => "SW_SHOW",
               6 => "SW_MINIMIZE",
               7 => "SW_SHOWMINNOACTIVE",
               8 => "SW_SHOWNA",
               9 => "SW_RESTORE",
              10 => "SHOWDEFAULT");
```

```perl
my %vol_type = (0 => "Unknown",
                 1 => "No root directory",
                 2 => "Removable",
                 3 => "Fixed",
                 4 => "Remote",
                 5 => "CD-ROM",
                 6 => "Ram drive");
# Get info about the file
#my ($size,$atime,$mtime,$ctime) = (stat($file))[7,8,9,10];
#print $file." $size bytes\n";
#print "Access Time      = ".gmtime($atime)." (UTC)\n";
#print "Creation Date    = ".gmtime($ctime)." (UTC)\n";
#print "Modification Time = ".gmtime($mtime)." (UTC)\n";
#print "\n";
# Open file in binary mode
open(FH,$file) || die "Could not open $file: $!\n";
binmode(FH);
seek(FH,$ofs,0);
read(FH,$record,0x4c);
if (unpack("Vx72",$record) == 0x4c) {
   my %hdr = parseHeader($record);
# print summary info from header
  print "Flags:\n";
  foreach my $i (keys %flags) {
  print $flags{$i}."\n" if ($hdr{flags} & $i);
  }
  print "\n";
  if (scalar keys %fileattr > 0) {
     print "Attributes:\n";
     foreach my $i (keys %fileattr) {
     print $fileattr{$i}."\n" if ($hdr{attr} & $i);
     }
     print "\n";
     }
     print "MAC Times: \n";
     print "Creation Time = ".gmtime($hdr{ctime})." (UTC)\n";
     print "Modification Time = ".gmtime($hdr{mtime})." (UTC)\n";
     print "Access Time = ".gmtime($hdr{atime})." (UTC)\n";
     print "\n";
     print "ShowWnd value(s):\n";
```

```perl
  foreach my $i (keys %showwnd) {
    print $showwnd{$i}."\n" if ($hdr{showwnd} & $i);
  }
  $ofs += 0x4c;
# Check to see if Shell Item ID List exists. If so, get the length
# and skip it.
  if ($hdr{flags} & 0x01) {
#      print "Shell Item ID List exists.\n";
    seek(FH,$ofs,0);
    read(FH,$record,2);
# Note: add 2 to the offset as the Shell Item ID list length is not
included in the
#      structure itself
    $ofs += unpack("v",$record) + 2;
  }
# Check File Location Info
  if ($hdr{flags} & 0x02) {
    seek(FH,$ofs,0);
    read(FH,$record,4);
    my $l = unpack("V",$record);
    if ($l > 0) {
      seek(FH,$ofs,0);
      read(FH,$record,0x1c);
      my %li = fileLocInfo($record);
    print "\n";
      if ($li{flags} & 0x1) {
# Get the local volume table
        print "Shortcut file is on a local volume.\n";
        my %lvt = localVolTable($ofs + $li{vol_ofs});
        print "Volume Name = $lvt{name}\n";
        print "Volume Type = ".$vol_type{$lvt{type}}."\n";
        printf "Volume SN = 0x%x\n",$lvt{vol_sn};
        print "\n";
      }
      if ($li{flags} & 0x2) {
# Get the network volume table
        print "File is on a network share.\n";
        my %nvt = netVolTable($ofs + $li{network_ofs});
        print "Network Share name = $nvt{name}\n";
      }
```

```perl
      if ($li{base_ofs} > 0) {
        my $basename = getBasePathName($ofs + $li{base_ofs});
        print "Base = $basename\n";
      }
      if ($li{path_ofs} > 0) {
        my $pathname = getPathName($ofs + $li{path_ofs});
        print "Path = ".$pathname."\n";
      }
    }
  }
}
else {
  die "$file does not have a valid shortcut header.\n"
}
close(FH);
sub parseHeader {
  my $data = $_[0];
  my %hdr;
  my @hd = unpack("Vx16V12x8",$data);
  $hdr{id}        = $hd[0];
  $hdr{flags}     = $hd[1];
  $hdr{attr}      = $hd[2];
  $hdr{ctime}     = getTime($hd[3],$hd[4]);
  $hdr{mtime}     = getTime($hd[5],$hd[6]);
  $hdr{atime}     = getTime($hd[7],$hd[8]);
  $hdr{length}    = $hd[9];
  $hdr{icon_num}  = $hd[10];
  $hdr{showwnd}   = $hd[11];
  $hdr{hotkey}    = $hd[12];
  undef @hd;
  return %hdr;
}
sub fileLocInfo {
  my $data = $_[0];
  my %fl;
  ($fl{len},$fl{ptr},$fl{flags},$fl{vol_ofs},$fl{base_ofs},$fl{network_ofs},
  $fl{path_ofs}) = unpack("V7",$data);
  return %fl;
}
sub localVolTable {
  my $offset = $_[0];
```

```perl
  my $data;
  my %lv;
  seek(FH,$offset,0);
  read(FH,$data,0x10);
  ($lv{len},$lv{type},$lv{vol_sn},$lv{ofs}) = unpack("V4",$data);
  seek(FH,$offset + $lv{ofs},0);
  read(FH,$data, $lv{len} - 0x10);
  $lv{name} = $data;
  return %lv;
}
sub getPathName {
  my $ofs = $_[0];
  my $data;
  my @char;
  my $len;
  my $tag = 1;
  while($tag) {
    seek(FH,$ofs,0);
    read(FH,$data,2);
    $tag = 0 if (unpack("v",$data) == 0x00);
    push(@char,$data);
    $ofs += 2;
  }
  return join('',@char);
}
sub getBasePathName {
  my $ofs = $_[0];
  my $data;
  my @char;
  my $len;
  my $tag = 1;
  while($tag) {
    seek(FH,$ofs,0);
    read(FH,$data,2);
    $tag = 0 if (unpack("v",$data) == 0x00);
    push(@char,$data);
    $ofs += 2;
  }
  return join('',@char);
}
```

```perl
sub netVolTable {
  my $offset = $_[0];
  my $data;
  my %nv;
  seek(FH,$offset,0);
  read(FH,$data,0x14);
  ($nv{len},$nv{ofs}) = unpack("Vx4Vx8",$data);
#     printf "Length of the network volume table = 0x%x\n",$nv{len};
#     printf "Offset to the network share name = 0x%x\n",$nv{ofs};
  seek(FH,$offset + $nv{ofs},0);
  read(FH,$data, $nv{len} - 0x14);
  $nv{name} = $data;
  return %nv;
}
sub getTime() {
  my $lo = shift;
  my $hi = shift;
  my $t;
  if ($lo == 0 && $hi == 0) {
    $t = 0;
  } else {
    $lo -= 0xd53e8000;
    $hi -= 0x019db1de;
    $t = int($hi`429.4967296 + $lo/1e7);
  };
  $t = 0 if ($t < 0);
  return $t;
}
```

Lslnk.pl will parse the contents of the Window shortcut file and display a great deal of information about the shortcut files, including the creation, last access and modification times not only of the shortcut file, but also those of the file that the shortcut points to, which is embedded within the binary content of the shortcut file.

# Registry

The Windows Registry is a binary hierarchal database that contains a great deal of valuable information for the analyst or investigator. According to Microsoft, the Registry holds configuration information for the system and applications, replacing the old initialization (*.ini, pronounced "eye-en-eye") files. For a forensic analyst, though, the

Registry can be looked at as one big log file. The Registry is made up of keys (the folders you see when you open up RegEdit), which contain subkeys and values, and values, which contain data. Figure II.1 illustrates keys, values, and data.

**Figure II.1** Extract from RegEdit showing keys, values, and data

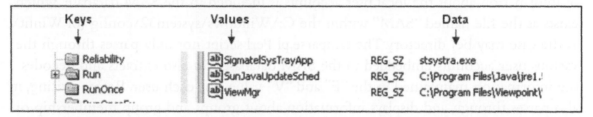

More information regarding the specific structure of the Registry, as well as its immense value in forensic analysis, please see my other book published by Syngress/ Elsevier, *Windows Forensic Analysis.*

As we saw in the previous Part, a great way to access the Registry on a live system is through the use of the Win32::TieRegistry module. However, when performing forensic analysis of Registry files extracted from an acquired image, the ideal module to use is the Parse::Win32Registry module,[2] from James McFarlane. I had looked into writing my own tools for accessing a raw Registry file and extracting keys and values, but while I was going about putting together the ground work for that module, James released a version of his Parse::Win32Registry module, which takes a completely object-oriented approach to the task. While we will be using James' module to parse Registry files from Windows 2000, XP, 2003, and yes, even Vista, Parse::Win32Registry can also be used to parse Registry files from Windows ME and 95/98, as well.

**NOTE**

As I'm writing this Part, the Parse::Win32Registry module is not available via the Perl Package Manager (ppm). In order to install this module, all you really need to do is download the .tar.gz file, decompress everything (WinZip works just fine), and copy the entire Parse folder to the \Perl\site\lib\ directory on your system.

---

[2] http://search.cpan.org/~jmacfarla/Parse-Win32Registry-0.30/

# SAMParse.pl

While a system is running, not even an administrator can access the SAM database, which contains information about the user accounts on the system (to include the password hashes). However, during a post-mortem analysis of the system, a great deal of information about the local user accounts is included in the SAM database, which exists as the file named "SAM" within the C:\Windows\system32\config (or Winnt\, as the case may be) directory. The samparse.pl Perl script not only parses through the various user accounts embedded in the SAM database, but also extracts and decodes the information maintained in the "F" and "V" values for each user. Before exiting, it also parses through and displays information about groups and group membership on the system (i.e., as it applies to local user accounts).

```
#! c:\perl\bin\perl.exe
#-------------------------------------------------------------
# samparse.pl
# Perl script to retrieve user information from a raw
# Registry/SAM file
#
# Usage:
# C:\Perl>samparse.pl <path_to_SAM_file> [> sam_user.txt]
#
# This script is intended to be used against SAM files
# extracted from
# from an image, either from the system32\config directory, or from system
# restore points.
#
# copyright 2006-2007 H. Carvey, keydet89@yahoo.com
#-------------------------------------------------------------
use strict;
use Parse::Win32Registry qw(:REG_);

# Included to permit compiling via Perl2Exe
#perl2exe_include "Parse/Win32Registry/Key.pm";
#perl2exe_include "Parse/Win32Registry/Value.pm";
my $sam = shift || die "You must enter a filename.\n";
die "$sam not found.\n" unless (-e $sam);
my %acb_flags = (0x0001 => "Account Disabled",
            0x0002 => "Home directory required",
         0x0004    => "Password not required",
         0x0008    => "Temporary duplicate account",
```

```perl
  0x0010 => "Normal user account",
  0x0020 => "MNS logon user account",
  0x0040 => "Interdomain trust account",
  0x0080 => "Workstation trust account",
  0x0100 => "Server trust account",
  0x0200 => "Password does not expire",
  0x0400 => "Account auto locked");
my $reg = Parse::Win32Registry->new($sam);
my $root_key = $reg->get_root_key;
my %users = getUsers();
print "-" x 25,"\n";
print "User Information\n";
print "-" x 25,"\n";
foreach my $rid (keys %users) {
  ($users{$rid}{fullname} eq "") ? (print $users{$rid}{name}."\n") :
    (print $users{$rid}{name}." (".$users{$rid}{fullname}.")\n");
  ($users{$rid}{comment} eq "") ? () : (print $users{$rid}{comment}."\n");
  print "Key LastWrite Time = ".gmtime($users{$rid}{lastwrite})." (UTC)\n";
  my $ll;
  ($users{$rid}{last_login} == 0)?($ll = "Never"):($ll = gmtime($users{$rid}{last_
login})." (UTC)");
  print "Last Login = ".$ll."\n";
  print "Login Count = ".$users{$rid}{login_count}."\n";
  my $prd;
  ($users{$rid}{pwd_reset_date} == 0)?($prd = "Never"):($prd =
gmtime($users{$rid}{pwd_reset_date})." (UTC)");
  print "Pwd Reset Date = ".$prd."\n";
  my $pfd;
  ($users{$rid}{pwd_fail_date} == 0)?($pfd = "Never"):($pfd =
gmtime($users{$rid}{pwd_fail_date})." (UTC)");
  print "Pwd Failure Date = ".$pfd."\n";
  print "Account Flags: \n";
  foreach my $flag (keys %acb_flags) {
    print " -> ".$acb_flags{$flag}."\n" if ($users{$rid}{flags} & $flag);
  }
  print "\n";
}
my %groups = getGroups();
print "-" x 25,"\n";
print "Group Information\n";
```

```perl
print "-" x 25,"\n";
foreach my $rid (keys %groups) {
  print $groups{$rid}{name}."\n";
  ($groups{$rid}{comment} eq "") ? () : (print $groups{$rid}{comment}."\n");
  print "Key LastWrite Time = ".gmtime($groups{$rid}{lastwrite})." (UTC)\n";
  my @users = split(/,/,$groups{$rid}{users});
  if ($groups{$rid}{users} eq "None") {
    print "\tNo Users\n";
  }
  else {
    foreach my $u (@users) {
      if (exists $users{$u}) {
        print "\t".$users{$u}{name}."\n";
      }
      else {
        print "\t$u\n";
      }
    }
  }
  print "\n";
}
sub getUsers {
  my %users = ();
  my $user_path = 'SAM\\Domains\\Account\\Users';
  my $users = $root_key->get_subkey($user_path);
  my @user_list = $users->get_list_of_subkeys();
  if (@user_list) {
    foreach my $u (@user_list) {
      my $rid = $u->get_name();
      my $ts = $u->get_timestamp();
      my $tag = "0000";
      if ($rid =~m/^$tag/) {
        my $v_value = $u->get_value("V");
        my $v = $v_value->get_data();
        my %v_val = parseV($v);
        $rid =~ s/^0000//;
        $rid = hex($rid);
        $users{$rid}{name} = $v_val{name};
        $users{$rid}{fullname} = $v_val{fullname};
        $users{$rid}{lastwrite} = $ts;
        $users{$rid}{comment} = $v_val{comment};
```

```perl
      my $f_value = $u->get_value("F");
      my $f = $f_value->get_data();
      my %f_val = parseF($f);
      $users{$rid}{last_login} = $f_val{last_login_date};
      $users{$rid}{pwd_reset_date} = $f_val{pwd_reset_date};
      $users{$rid}{pwd_fail_date} = $f_val{pwd_fail_date};
      $users{$rid}{flags} = $f_val{acb_flags};
      $users{$rid}{login_count} = $f_val{login_count}
    }
   }
  }
  else {
   undef %users;
  }
  return %users;
}
sub getGroups {
  my %sam_groups = ();
  my $grppath = 'SAM\\Domains\\Builtin\\Aliases';
  my $groups = $root_key->get_subkey($grppath);
  my %grps;
  foreach my $k ($groups->get_list_of_subkeys()) {
    if ($k->get_name() = ~ m/^0000/) {
      $grps{$k->get_name()}{LastWrite} = $k->get_timestamp();
      $grps{$k->get_name()}{C_value} = $k->get_value("C")->get_data();
    }
  }
  foreach my $k (keys %grps) {
    my $name = $k;
    $name =~s/^0000//;
    $sam_groups{$name}{lastwrite} = $grps{$k}{LastWrite};
    my %c_val = parseC($grps{$k}{C_value});
    $sam_groups{$name}{name} = $c_val{group_name};
    $sam_groups{$name}{comment} = $c_val{comment};
    $sam_groups{$name}{users} = $c_val{users};
  }
  return %sam_groups;
}
sub parseF {
  my $f = shift;
  my %f_value = ();
```

```perl
# last login date
  $f_value{last_login_date} = _getTimeDate(unpack("VV",substr($f,8,8)));
#      password reset/acct creation
  $f_value{pwd_reset_date} = _getTimeDate(unpack("VV",substr($f,24,8)));
# Account expires
  $f_value{acct_exp_date} = _getTimeDate(unpack("VV",substr($f,32,8)));
# Incorrect password
  $f_value{pwd_fail_date} = _getTimeDate(unpack("VV",substr($f,40,8)));
  $f_value{rid} = unpack("V",substr($f,48,4));
  $f_value{acb_flags} = unpack("v",substr($f,56,2));
  $f_value{failed_count} = unpack("v",substr($f,64,2));
  $f_value{login_count} = unpack("v",substr($f,66,2));
  return %f_value;
}

sub parseV {
  my $v = shift;
  my %v_val = ();
  my $header = substr($v,0,44);
  my @vals = unpack("V*",$header);
  $v_val{name} = _uniToAscii(substr($v,($vals[3] + 0xCC),$vals[4]));
  $v_val{fullname} = _uniToAscii(substr($v,($vals[6] + 0xCC),$vals[7]))
if ($vals[7] > 0);
  $v_val{comment} = _uniToAscii(substr($v,($vals[9] + 0xCC),$vals[10]))
if ($vals[10] > 0);
  return %v_val;
}

sub parseC {
  my $cv = $_[0];
  my %c_val = ();
  my $header = substr($cv,0,0x34);
  my @vals = unpack("V*",$header);
  $c_val{group_name} = _uniToAscii(substr($cv,(0x34 + $vals[4]),$vals[5]));
  $c_val{comment} = _uniToAscii(substr($cv,(0x34 + $vals[7]),$vals[8]));
  my $num = $vals[12];
  my @users = ();
  my $ofs;
  $num -= 2 if ($c_val{group_name} eq "Users");
  if ($num > 0) {
    my $count = 0;
    foreach my $c (1..$num) {
```

```perl
        $ofs = ($vals[10] + 52 + 25 + $count - 1);
        $ofs = ($vals[10] + 52 + 25 + $count - 1 + 24) if ($c_val{group_name}
eq "Users");
        my $rid = unpack("v",substr($cv,$ofs,2));
        push(@users,$rid);
        $count += (27 + 1) if ($count < $vals[11]);
      }
    }
    if ((scalar @users) > 0) {
      $c_val{users} = join(',',@users);
    }
    else {
      $c_val{users} = "None";
    }
  return %c_val;
}
#---------------------------------------------------------------
# _getTimeDate()
# Input : 2 DWORDs, each containing half of the LastWrite time
# Output: readable GMT time string
#---------------------------------------------------------------
sub _getTimeDate {
# Borrowed from Andreas Schuster's ptfinder code
  my $lo = shift;
  my $hi = shift;
  my $t;

  if ($lo == 0 && $hi == 0) {
    $t = 0;
  } else {
    $lo -= 0xd53e8000;
    $hi -= 0x019db1de;
    $t = int($hi*429.4967296 + $lo/1e7);
  };
  $t = 0 if ($t < 0);
  return $t;
}
sub _uniToAscii {
my $str = $_[0];
$str =Ð s/\00//g;
return $str;
}
```

When samparse.pl is run against a SAM database file extracted from an acquired image, the investigator is presented with information such as what appears below:

```
--------------------------------------------------------------
User Information
--------------------------------------------------------------
SUPPORT_388945a0 (CN=Microsoft Corporation,L=Redmond,S=Washington,C=US)
This is a vendor's account for the Help and Support Service
Key LastWrite Time = Wed Aug 18 00:39:28 2004 (UTC)
Last Login         = Never
Login Count        = 0
Pwd Reset Date     = Wed Aug 18 00:39:27 2004 (UTC)
Pwd Failure Date   = Never
Account Flags:
  --> Password does not expire
  --> Account Disabled
  --> Normal user account
Guest
Built-in account for guest access to the computer/domain
Key LastWrite Time  = Tue Aug 17 20:27:13 2004 (UTC)
Last Login          = Never
Login Count         = 0
Pwd Reset Date      = Never
Pwd Failure Date    = Never
Account Flags:
  --> Password does not expire
  --> Account Disabled
  --> Password not required
  --> Normal user account

jdoe (John Doe)
Corporate User
Key LastWrite Time   = Mon Sep 26 22:55:51 2005 (UTC)
Last Login           = Mon Sep 26 22:55:51 2005 (UTC)
Login Count          = 2
Pwd Reset Date       = Fri Sep 9 01:09:49 2005 (UTC)
Pwd Failure Date     = Mon Sep 26 22:55:49 2005 (UTC)
Account Flags:
  --> Password does not expire
  --> Normal user account

HelpAssistant (Remote Desktop Help Assistant Account)
Account for Providing Remote Assistance
```

```
Key LastWrite Time      = Wed Aug 18 00:37:19 2004 (UTC)
Last Login              = Never
Login Count             = 0
Pwd Reset Date          = Wed Aug 18 00:37:19 2004 (UTC)
Pwd Failure Date        = Never
Account Flags:
  --> Password does not expire
  --> Account Disabled
  --> Normal user account
Administrator
Built-in account for administering the computer/domain
Key LastWrite Time   = Tue Aug 17 20:31:47 2004 (UTC)
Last Login     = Never
Login Count    = 0
Pwd Reset Date         = Tue Aug 17 20:31:47 2004 (UTC)
Pwd Failure Date       = Never
Account Flags:
  --> Password does not expire
  --> Normal user account
Harlan
Key LastWrite Time   = Mon Sep 26 23:37:51 2005 (UTC)
Last Login           = Mon Sep 26 23:37:51 2005 (UTC)
Login Count          = 35
Pwd Reset Date       = Wed Aug 18 00:49:42 2004 (UTC)
Pwd Failure Date     = Mon Sep 26 23:37:47 2005 (UTC)
Account Flags:
  --> Password does not expire

  --> Normal user account

------------------------------------------------------------

Group Information
------------------------------------------------------------

Users
Users are prevented from making accidental or intentional system-wide changes.
Thus, Users can run certified applications, but not most legacy applications
Key LastWrite Time   = Fri Sep 9 01:09:49 2005 (UTC)
   jdoe

Network Configuration Operators
Members in this group can have some administrative privileges to manage configur
ation of networking features
Key LastWrite Time   = Tue Aug 17 20:27:13 2004 (UTC)
   No Users
```

```
Backup Operators
Backup Operators can override security restrictions for the sole purpose of
backing up or restoring files
Key LastWrite Time   = Tue Aug 17 20:27:13 2004 (UTC)
    No Users

Replicator
Supports file replication in a domain
Key LastWrite Time   = Tue Aug 17 20:27:13 2004 (UTC)
    No Users

Administrators
Administrators have complete and unrestricted access to the computer/domain
Key LastWrite Time   = Wed Aug 18 00:46:24 2004 (UTC)
    Administrator
    Harlan

Power Users
Power Users possess most administrative powers with some restrictions. Thus, Power
Users can run legacy applications in addition to certified applications
Key LastWrite Time   = Tue Aug 17 20:27:13 2004 (UTC)
    No Users

Guests
Guests have the same access as members of the Users group by default, except for
the Guest account which is further restricted
Key LastWrite Time   = Tue Aug 17 20:27:13 2004 (UTC)
    Guest

Remote Desktop Users
Members in this group are granted the right to logon remotely
Key LastWrite Time   = Tue Aug 17 20:27:13 2004 (UTC)
    No Users
```

As you can see, samparse.pl displays local user account information based on what is extracted from the SAM database. Within the "Group Membership" section, you may see series of 4 or 5 digits where you would expect to see a user name, such as beneath the Administrators group (above). In such cases, these values are relative identifiers (RIDs) for domain users. This is an indication that the system being examined is (or was, at one point) part of a domain.

# SECParse.pl

Like the SAM file within the system32\config directory, the Security file also contains some very useful information. In particular, the Security file maintains the audit configuration (as seen when running auditpol.exe on a live system) which determines which events are audited and recorded in the Windows Event Log. Microsoft Knowledge Base article Q246120[3] describes how to parse this information from the Security file extracted from a Windows NT 4.0 system, and sources available on the Internet provide the necessary information in order to allow you to do the same thing for Windows 2000, XP, and 2003 systems.

```perl
#! c:\perl\bin\perl.exe
#----------------------------------------------------------------
# secparse.pl
# Parse the raw Security file and display the audit policy
# for the system (2000, XP, 2003)
#
# Usage: secparse.pl <filename> [ > output_file]
#
# NT: http://support.microsoft.com/kb/246120
# 2K: http://www.jsifaq.com/SF/Tips/Tip.aspx?id=5231
# http://loguk.blogspot.com/2004/09/bit-of-windows-internals-recently.html
#
# copyright 2006-2007 H. Carvey keydet89@yahoo.com
#----------------------------------------------------------------
use strict;
use Parse::Win32Registry qw(:REG_);
my %win2kevents = (0 => "System Events",
                   1 => "Logon Events",
                   2 => "Object Access",
                   3 => "Privilege Use",
                   4 => "Process Tracking",
                   5 => "Policy Change",
                   6 => "Account Management",
                   7 => "Directory Service Access",
                   8 => "Account Logon Events");
my %ntevents = (0 => "Restart, Shutdown, Sys",
                1 => "Logon/Logoff",
                2 => "File/Object Access",
                3 => "Use of User Rights",
                4 => "Process Tracking",
                5 => "Sec Policy Mgmt",
                6 => "User/Grp Mgmt");
```

---

[3] http://support.microsoft.com/default.aspx?scid=kb;EN-US;q246120

```perl
my %audit = (0 => "None",
             1 => "Succ",
             2 => "Fail",
             3 => "Both");
my %policy = ();

my $file = shift || die "You must enter a filename.\n";
die "$file not found.\n" unless (-e $file);

my $reg = Parse::Win32Registry->new($file);
my $root = $reg->get_root_key;
my $pol = $root->get_subkey("Policy\\PolAdtEv");
my $ts = $pol->get_timestamp();
print "LastWrite: ".gmtime($ts)." (UTC)\n";

my $val = $pol->get_value("");
#print "\t".$val->print_summary()."\n";
my $adt = $val->get_data();
my $len = length($adt);

my $enabled = unpack("C",substr($adt,0,1));
if ($enabled) {
  print "Auditing was enabled.\n";
    my @evts = unpack("V*",substr($adt,4,$len-4));
    my $tot = $evts[scalar(@evts) - 1];
    print "There are $tot audit categories.\n";
  print "\n";

  if ($tot == 9) {
     foreach my $n (0..(scalar(@evts) - 2)) {
        my $adtev = $audit{$evts[$n]};
        $policy{$win2kevents{$n}} = $adtev;
     }
  }
  elsif ($tot == 7) {
    foreach my $n (0..(scalar(@evts) - 2)) {
        my $adtev = $audit{$evts[$n]};
        $policy{$ntevents{$n}} = $adtev;
    }
  }
  else {
    print "Unknown audit configuration.\n";
  }
```

```
    foreach my $k (keys %policy) {
      printf "%-25s %-4s\n",$k,$policy{$k};
    }
}
else {
  print "Auditing was not enabled.\n";
}
```

In order to run the secparse.pl Perl script, simply open a command prompt to the directory where the script is stored and type the following command:

```
C:\Perl\forensics>secparse.pl d:\cases\case001\security
```

When running secparse.pl against a Security file extracted from an acquired image, the investigator should expect to see something similar to the following:

```
LastWrite: Fri Sep 9 01:11:43 2005 (UTC)
Auditing was enabled.
There are 9 audit categories.
Privilege Use                 None
Object Access                 None
Account Logon Events          Both
System Events                 Both
Policy Change                 Both
Logon Events                  Both
Account Management            Both
Directory Service Access      None
Process Tracking              None
```

You'll notice that one of the first things returned by the secparse.pl Perl script is the LastWrite time of the PolAdtEv Registry key. I opted to do this so that the examiner can get an idea of when the audit configuration on the system may have been changed in some way. This may be extremely valuable information during an investigation, particularly when correlated with additional data from other sources.

As you will see in the "Event Log" section further on in this book, information derived from the use of secparse.pl Perl script can be extremely useful in an investigation, as it will inform the examiner of what she should expect to see in the various Event Log files.

# Recentdocs.pl

Another important aspect of the Registry can be found in the user's profile, specifically in the user's NTUSER.DAT file. This file maintains a great deal of useful information about the activities performed by whoever logged into the system using the user's credentials (i.e., username and password). For example, when the user opens various files on the system, not only are Windows shortcut files created, but so are entries within the user's RecentDocs Registry key (in the NTUSER.DAT file). Some of this information is in easily searched and viewed ASCII format, but other information is stored in a binary format that must be parsed in order to be understandable and therefore useable by the examiner. The recentdocs.pl uses the Parse::Win32Registry module to parse this information for easy display and viewing.

```perl
#! c:\perl\bin\perl.exe
--------------------------------------------------------------
# recentdocs.pl
#
# Parse the contents of the RecentDocs Registry key from a
# user's NTUSER.DAT
# file; all information is sent to STDOUT
#
# Usage
# C:\perl>u_recentdocs.pl <file>
# - or, create a harness file that calls the file via a
# 'require' pragma,
#       and launches getRecentDocs() by passing the filename (or
# use backticks,
#       and capture output to a list)
#
# ChangeLog
#       20070703 - Created
#
# copyright 2007 H. Carvey, keydet89@yahoo.com
--------------------------------------------------------------
use strict;
use Parse::Win32Registry qw(:REG_);

my $ntuser = shift || die "You must enter a filename.\n";
die "$ntuser not found.\n" unless (-e $ntuser);

\getRecentDocs($ntuser);

sub getRecentDocs {
  my $reg = Parse::Win32Registry->new($ntuser);
  my $root_key = $reg->get_root_key;
#print "Root key: $root_key\n";
```

```perl
   my $key_path = "Software\\Microsoft\\Windows\\CurrentVersion\\
Explorer\\RecentDocs";
   my $r_docs = $root_key->get_subkey($key_path);
   if (!defined($r_docs)) {
     print $key_path." not found.\n";

   }
   else {
# RecentDocs key exists, so get the values within the key, and
# the values within each subkey.
     my $ts = $r_docs->get_timestamp();
     my $main_name = $r_docs->get_name();
     print $main_name." [".gmtime($ts)." (UTC)]\n";

     my %vals = getValues($r_docs);
     if (scalar(keys %vals) < 2) {

     }
     else {
       foreach my $v (sort {$a <=> $b} keys %vals) {
         printf "%9s %-30s\n",$v,$vals{$v} unless ($v eq "MRUListEx");
       }
       printf "%9s %-30s\n", "MRUListEx",$vals{MRUListEx};
       print "\n";
       my @r_values = $r_docs->get_list_of_subkeys();
       foreach my $r (@r_values) {
         $ts = $r->get_timestamp();
         print $main_name."\\".$r->get_name()." [".gmtime($ts)." (UTC)]\n";

         %vals = getValues($r_docs);
         foreach my $v (sort {$a <=> $b} keys %vals) {
           printf "%9s %-30s\n",$v,$vals{$v} unless ($v eq "MRUListEx");
         }
         printf "%9s %-30s\n", "MRUListEx",$vals{MRUListEx};
         print "\n";
       }
     }
   }
}
# Get the values in a key
sub getValues {
  my $key_path = shift;
  my %key_values;
  my @vals = $key_path->get_list_of_values();
  foreach my $v (@vals) {
```

```perl
        $key_values{$v->get_name()} = $v->get_data();
    }
    return sortValues(%key_values);
}
# Sort the values in a hash
sub sortValues {
  my %v = @_;
  my %sorted;
  my $mru = "MRUList";
  foreach my $i (keys %v) {
    if ($i = ~ m/^$mru/i) {
      if (length($v{$i}) > 4) {
        my @mru = unpack("V*",$v{$i});
        pop(@mru);
        my $str = join(',',@mru);
        $sorted{$i} = $str;
      }
    }
    else {
      my $str = parseBinary($v{$i});
      $sorted{$i} = $str;
    }
  }
  return %sorted;
}
# Return a null-terminated string from within a binary data
# type
sub parseBinary {
  my $binary = shift;
  my @list = unpack("v*",$binary);
  my $count = 0;
  my $tag = 1;
  my $str;

  while($tag) {
    if ($list[$count] == 0) {
      $tag = 0;
    }
    else {
      my $i = $list[$count];
      $i = ~ s/\00//;
      $str .= pack("C",$i);
```

```
    }
    $count++;
  }
  return $str;
}
1;
```

Running the recentdocs.pl Perl script is fairly straightforward:

```
C:\Perl\forensics>recentdocs.pl d:\cases\case007\ntuser.dat
```

An excerpt of the output of the recentdocs.pl Perl script looks like:

```
RecentDocs [Mon Sep 26 23:33:07 2005 (UTC)]
   0 britney.jpg
   1 jdoe
   2 lads.zip
   3 hand1.gif
   4 Search Results
   5 hand2.gif
   6 alicia.silverstone.jpg
   7 LADS_ReadMe.txt
   8 010219_2100 (D:)
   9 README.TXT
  10 trout.ini
  11 small.gif
  12 honeynet_papers
  13 cover.jpg
  14 USB DISK (E:)
  15 fspconfig.jpg
  16 fru.jpg
  17 test.txt
  18 c$ on '192.168.1.22' (Z:)
  19 2k3_usb.log
  20 c$ on '192.168.1.71' (X:)
MRUListEx 20,19,18,17,14,16,15,13,12,11,8,10,9,7,1,6,0,4,5,3,2
```

As with any other Perl script, the output is easily formatted for display by the examiner…with some simple Perl code, the output can be displayed in ASCII format to standard output (as above), or to just about any other format and location, such as to a spreadsheet or database, for example.

# UAssist.pl

As mentioned in the previous Part, the UserAssist key within the user's NTUSER. DAT file contains a great deal of information that can be extremely valuable to an examiner. I've used the information derived from this key (actually, from its subkeys) in order to demonstrate that a user did, in fact, install and launch an application that had been uninstalled and removed prior to my arrival, as well as to show that a particular user account had access to the Windows shell (i.e., Windows Explorer) via the Remote Desktop client.

As you're undoubtedly aware by now, accessing the contents of the UserAssist key is a bit different when done in a post-mortem examination vice on a live system, but some of the other core functionality of the uassist.pl script (below) is no different from the Perl script by the same name listed in Part I.

```perl
#! c:\perl\bin\perl.exe
-------------------------------------------------------------
# parse NTUSER.DAT file, and list the contents of one of the
# UserAssist\GUID\Count keys, sorted by most recent time
#
# Usage
# C:\perl>u_uassist.pl <file>
# - or, create a harness file that calls the file via a
# 'require' pragma,
#      and launches getUserAssist() by passing the filename (or
# use backticks,
#      and capture output to a list)
#
# ChangeLog
# 20070703 - Created
#
# copyright 2007 H. Carvey
-------------------------------------------------------------
use strict;
use Parse::Win32Registry qw(:REG_);
# Included to permit compiling via Perl2Exe
#perl2exe_include "Parse/Win32Registry/Key.pm";
#perl2exe_include "Parse/Win32Registry/Value.pm";
my $ntuser = shift || die "You must enter a filename.\n";
die "$ntuser not found.\n" unless (-e $ntuser);

\getUserAssist($ntuser);
# \getUserAssist($ntuser,0);
sub getUserAssist {
  my $ntuser = shift;
# Two levels for output format
```

```perl
# 0 = All entries, plus time-sorted values
# 1 = only time-sorted values       (default)
  my $level = shift;
  $level = 1 if (!(defined($level)));
  my $reg = Parse::Win32Registry->new($ntuser);
  my $root_key = $reg->get_root_key;
# The first thing we want to do is check and see if the Settings subkey
# exists, and if so, have values such as NoLog and NoEncrypt been set
  my $settings_path = 'Software\\Microsoft\\Windows\\CurrentVersion\\Explorer\\'.
        'UserAssist\\Settings';
  my $settings;
  if ($settings = $root_key->get_subkey($settings_path)) {
    print "Settings subkey [".gmtime($settings->get_timestamp())." (UTC)]\n";
    my @settings_values = $settings->get_list_of_values();
    if (scalar(@settings_values) > 0) {
      foreach my $v (@settings_values) {
        printf "%-10s %-10s\n",$v->get_name(),$v->get_data();
      }
      print "\n";
    }
    else {
      print "No values found.\n\n";
    }
  }
  else {
    print "UserAssist\\Settings subkey not found.\n\n";
  }
#print "Root key: $root_key\n";
  my $key_path = 'Software\\Microsoft\\Windows\\CurrentVersion\\Explorer\\
UserAssist\\'.
      '{75048700-EF1F-11D0-9888-006097DEACF9}\\Count';
  my $hrzr = "HRZR";
  my $ueme = "UEME";
  my $count = $root_key->get_subkey($key_path);
  print "UserAssist (Active Desktop) [".gmtime($count->get_timestamp())."
(UTC)]\n";
  my %ua = ();
  foreach my $value ($count->get_list_of_values) {
    my $value_name = $value->get_name();
    my $data = $value->get_data();
```

```perl
    my ($freq,$val1,$val2) = unpack("x4VVV",$data);
    if (length($data) == 16 && $val2 != 0) {
      my $time_value = getTime($val1,$val2);
# Check the value name to see if it begins with "HRZR"; this
# indicates ROT-13
# encryption; if so, decrypt. If NoEncrypt had been set after
# some of the values
# had been written to the Count key, then the decryption
# routine would work on some
# values and not others.
        if ($value_name = ~ m/^$hrzr/) {
          $value_name = ~ tr/N-ZA-Mn-za-m/A-Za-z/;
        }
        print $value_name."\n" if ($level == 0);
  push(@{$ua{$time_value}},$value_name.";".$freq);
    }
  }
  print "\n" if ($level == 0);
  foreach my $t (reverse sort {$a <=> $b} keys %ua) {
    print gmtime($t)." (UTC)\n";
    foreach my $item (@{$ua{$t}}) {
      print "\t$item\n";
    }
  }
}
#-------------------------------------------------------------
# getTime()
# Translate FILETIME object (2 DWORDS) to Unix time, to be
# passed to gmtime() or localtime()
#-------------------------------------------------------------
sub getTime() {
  my $lo = shift;
  my $hi = shift;
  my $t;
  if ($lo == 0 && $hi == 0) {
    $t = 0;
  } else {
    $lo -= 0xd53e8000;
    $hi -= 0x019db1de;
    $t = int($hi*429.4967296 + $lo/1e7);
```

```
    };
    $t = 0 if ($t < 0);
    return $t;
}
1;
```

Figure II.2 illustrates an excerpt of the output of the uassist.pl Perl script when run against an NTUSER.DAT file, using a command similar to the following:

```
C:\Perl\forensics>uassist.pl d:\cases\case007\ntuser.dat
```

One of the interesting aspects of this script is that it maintains a hash of lists (see the Perl docs under "perldsc")[4] of the value data that contains timestamps, and then sorts them in reverse order, so that all of the user activity that occurred via the Windows shell and was recorded in the key can be displayed in order, with the most recent time displayed first. The key line within the Perl script itself that allows this to occur is:

```
foreach my $t (reverse sort {$a <=> $b} keys %ua) {
```

Pretty interesting, isn't it? Perhaps not if you're an extremely experienced Perl programmer, but when you're a forensic examiner interested in establishing a timeline, having the information you need displayed in an easy-to-understand format (albeit in UTC time format) can be a blessing!

**Figure II.2** Output of uassist.pl script

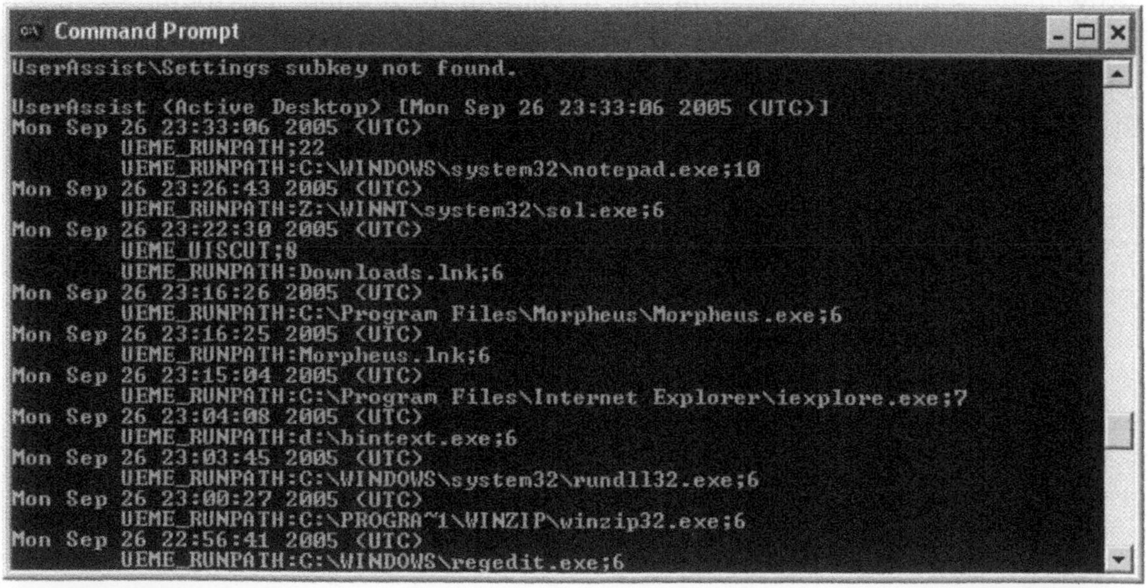

---

[4] http://www.perl.com/doc/FMTEYEWTK/pdsc/pdsc-2.html

# Event Logs

Many times, an examination of the Windows Event Log event records will provide some very useful information that may affect your investigation. The Event Log is capable of holding a fairly amazing array of information, from records of failed attempts to login into the system to the system being shutdown and rebooted. When working with the Event Log on a live system, most folks will interface with it through the EventViewer. One of the techniques that a forensic analyst may use to analyze an Event Log during an investigation is to extract the file from within the image and then attempt to open the log in the Event Viewer on their analysis system. However, this does not always work... many analysts have reported receiving an error message stating that the Event Log is "corrupt". This message is reported by the Windows API – so what if, like the Registry, we can parse the contents of the Event Log files without using the API?

# Evt2xls.pl

Evt2xls.pl is a Perl script that I developed over time, and have found it to be extremely useful. I started with a simple script that parsed through the .evt file in binary mode, and retrieved event records for me, writing them out to the console. This got to be somewhat cumbersome over time, as there was more that I wanted to do with the script, so I wrote a module to encapsulate and hide the routines I'd been writing. From there, I would have the output sent to the console as comma-separated values, after which I would redirect the output to a file and then open that file in Excel. However, the date format was never write, and sometimes the messages would have commas... both of these threw off my analysis. So, I added some code to write the event record entries directly to a spreadsheet that is binary compatible with MS Excel (via the Spreadsheet::WriteExcel module) and also added some code to do some frequency analysis of event IDs.

Evt2xls.pl appears as follows (and is available on the accompanying DVD):

```
#! c:\perl\bin\perl.exe
#-----------------------------------------------------------
# evt2xls.pl, version 20070611
# Parse Windows 2000, XP, 2003 EventLog files in binary format,
# putting the event
# records into an Excel spreadsheet; can also generate a report
# showing event
# source/ID frequencies (for Security Event Log, login type is
```

```perl
# added to the
# event ID), suitable for entry into eventid.net
#
# see _syntax() usage for examples
#
# ChangeLog:
#      20070611: created
#
# copyright 2007 H. Carvey, keydet89@yahoo.com
#----------------------------------------------------------------
use strict;
use ReadEvt;
use Spreadsheet::WriteExcel;
use Getopt::Long;

my %config;
Getopt::Long::Configure("prefix_pattern=(-|\/)");
GetOptions(\%config,qw(event|e=s output|o=s report|r=s help|?|h));
if ($config{help} || !%config) {
  _syntax();
}

die "No Event Log file name entered.\n" if (! $config{event});
die "No output spreadsheet file name entered.\n" if (! $config{output});

my $file = $config{event};
my $reportfile = $config{report};
my $login = $config{login};
my $outfile = $config{output};

die "$file not found.\n" unless (-e $file);
my $name;
# Parse the filename
if (grep(/\\/,$file)) {
  my @vals = split(/\\/,$file);
  my $i = scalar(@vals) - 1;
  $name = (split(/\./,$vals[$i]))[0];
}
else {
# No path separators, so the file may be stored in the same directory
  $name = split(/\./,$file);
}

my $evt;
if ($evt = ReadEvt::new($file)) {
```

```
#       print "EVT Object created.\n";
}
else {
  print "EVT Object not created. Exiting.\n";
  exit 1;
}
my %hdr = ();
if (%hdr = $evt->parseHeader()) {
# no need to do anything...
#       print "Header parsed...\n";
}
else {
  print "Error : ".$evt->getError()."\n";
  die;
}
# Set up to generate a report;
my $total = 0;
my %er;
my %dates;
my $wb = Spreadsheet::WriteExcel->new($outfile);
my $format = $wb->add_format();
$format->set_num_format('mmm d yyyy hh:mm AM/PM');
# Add a worksheet
my $ws = $wb->add_worksheet($name);
my $row = 0;
$ws->write($row, 0, "Record Number");
$ws->write($row, 1, "Source");
$ws->write($row, 2, "ComputerName");
$ws->write($row, 3, "Event ID");
$ws->write($row, 4, "Event Type");
$ws->write($row, 5, "Time Generated");
$ws->write($row, 6, "User SID");
$ws->write($row, 7, "Strings");

my $ofs = $evt->getFirstRecordOffset();

while ($ofs) {
  $row++;
  my %record = $evt->readEventRecord($ofs);
  my $time_gen = dateConvert($record{time_gen});
  $ws->write($row, 0, $record{rec_num});
  $ws->write($row, 1, $record{source});
```

```perl
  $ws->write($row, 2, $record{computername});
  $ws->write($row, 3, $record{evt_id});
  $ws->write($row, 4, $record{evt_type});
  $ws->write($row, 5, $time_gen , $format);
  $ws->write($row, 6, $record{sid});
  $ws->write($row, 7, $record{strings}) if ($record{num_str} > 0);
# Only collect report stats if necessary
  if ($config{report}) {
    $total++;
# If the Security Event Log is being parsed, add the login type
to the specific login IDs
    if ($record{source} eq "Security" && ($record{evt_id} > 527 && $record{evt_id}
< 541)) {
      my $type = getLoginType($record{evt_id},$record{strings});
      $record{evt_id} = $record{evt_id}.",".$type;
    }
    if (exists $er{$record{source}.":".$record{evt_id}}) {
      $er{$record{source}.":".$record{evt_id}}++;
    }
    else {
      $er{$record{source}.":".$record{evt_id}} = 1;
    }
    $dates{$record{time_gen}} = 1;
  }
# length of record is $record{length}...skip forward that far
  $ofs = $evt->locateNextRecord($record{length});
#      printf "Current Offset = 0x%x\n",$evt->getCurrOfs();
}
$evt->close();

# Generate the report
if ($config{report}) {
  open(RPT,">",$reportfile) || die "Could not open $reportfile: $!\n";
  print RPT "From the Event Log header: \n";
  print RPT "Oldest ID : ".$hdr{oldestID}."\n";
  print RPT "Next ID : ".$hdr{nextID}."\n";
  print RPT "Total Events : ".($hdr{nextID} - $hdr{oldestID})."\n";
  print RPT "-" x 30,"\n";
  print RPT "Total number of events counted: ".$total."\n";
  print RPT "-" x 30,"\n";
  print RPT "Event Source/ID Frequency\n";
  print RPT "\n";
```

```perl
      printf RPT "%-40s %10s %8s\n","Source","Event ID","Count";
      printf RPT "%-40s %10s %8s\n","-" x 10,"-" x 8,"-" x 5;
      my $er_total = 0;
      foreach my $i (sort keys %er) {
        my ($source,$id) = split(/:/,$i,2);
        printf RPT "%-40s %10s %8s\n",$source,$id,$er{$i};
        $er_total += $er{$i};
      }
      print RPT "\n";
      print RPT "Total: ".$er_total."\n";
      print RPT "\n";
      print RPT "-" x 30,"\n";
      print RPT "Date Range, in UTC\n";
      my @daterange = sort {$a <=> $b} keys %dates;
      my $i = scalar(@daterange) - 1;
      print RPT gmtime($daterange[0])." to ".gmtime($daterange[$i])."\n";
      close(RPT);
}

sub dateConvert {
  my $input = shift;
# Divide timestamp by number of seconds in a day.
# This gives a date serial with '0' on 1 Jan 1970.
  my $serial = $input / 86400;
  $serial += 25569;
  return $serial;
}

sub getLoginType {
  my $id = shift;
  my $strings = shift;
  my @vals = split(/\00/,$strings);
  if ($id == 528 || $id == 538 || $id == 540) {
    return $vals[3];
  }
  elsif ($id == 529 || $id = ~ m/^53/) {
    return $vals[2];
  }
  else {
    return 0;
  }
}
```

```
sub _syntax {
  print<< "EOT";
Evt2XLS [-e eventlog_file] [-o output_spreadsheet] [-r report_file] [-h]
Parse Windows 2000, XP, 2003 EventLog files in binary mode, converting to
binary Excel spreadsheet format; can also generate reports/stats (contains
event source/ID frequency info)
  -e eventlog_file............EventLog file to parse
  -o output_spreadsheet......spreadsheet file name to create
  -r report_file.............name of file to create report in
  -h.........................Help (print this information)
Ex: C:\\>evt2xls -e secevent.evt -o secevent.xls
    C:\\>evt2xls -e appevent.evt -o appevent.xls -r app_stats.log
copyright 2007 H. Carvey
EOT
}
```

Something to note about evt2xls.pl is that it uses a module called ReadEvt.pm
(which is included on the accompanying media along with the Perl script). I wrote
this module to encapsulate the code for parsing the Windows 2000, XP and 2003
binary Event Logs (.evt files). Installation of this module involves nothing more than
copying it into the same directory as the evt2xls.pl Perl script.

Evt2xls.pl also relies upon (or "uses") two other modules, as well…Spreadsheet::
WriteExcel and Getopt::Long. Both of these modules can be easily installed via the
Perl Package Manager on the ActiveState installation of Perl, using the following
commands:

```
C:\perl>ppm install spreadsheet-writeexcel
```
       …and…
```
C:\perl>ppm install getopt-long
```

---

**NOTE**

Evt2xls.pl only works on Event Logs from Windows 2000, XP and 2003
systems. With Vista, Microsoft changed many things about the Event Log, to
include the structure and format of the files themselves. Early in 2007,
Andreas Schuster did considerable work in examining and parsing[5] these files,
providing (you guessed it!) a Perl script to parse the Vista and Windows 2008
Event Logs into plain text.

---

[5] http://computer.forensikblog.de/en/2007/08/evtx_parser.html

Let's look at an example of how to launch evt2xls.pl and parse a Windows Event Log file. To do that, let's assume that we've got an external hard drive attached to our analysis system via USB, and that drive has been mounted as F:\. On that external hard drive is a directory called "cases", which contains the Security, Application, and System Event Log files from a Windows system. In order to use evt2xls.pl, we would need to type in a command line such as:

```
C:\perl\forensics>evt2xls.pl -e f:\cases\secevent.evt -o
f:\cases\secevt.xls -r f:\cases\secevt.rpt
```

The "-e" switch tells the script which Event Log file to open, and the "-o" switch gives the script the path and filename of the Excel spreadsheet where the parsed event records will be written. The resulting output file is a binary spreadsheet file, which can be easily opened in Excel. The dates listed in the spreadsheet (i.e., the dates/times that the events were generated, which is part of the event record structure) have been converted to a format that Excel understands, allowing you to sort the spreadsheet based on any of the visible columns. Finally, the "-r" switch tells the script were to write the report file, which contains information such as the frequency with which each event ID occurs within the Event Log file, as the well as the date ranges of the event records. When the spreadsheet and report file are combined with sources that describe why each event record is generated (such as http://www.eventid.net), they provide a powerful set of analysis tools for the forensic examiner.

## Swiss Army Knife

### Extending evt2xls.pl

Forensic analysts not too terribly familiar with command line tools may find evt2xls.pl a little cumbersome to use, due to the fact that they need to enter the path for the various options. One way to extend the use of the script and make it a bit easier to use would be to simply provide a path for an output directory for the spreadsheet and the report. Many times when I'm performing analysis, the images themselves are located on an external hard drive, and I may not want to store the reports in that location. Other times, I have received

Event Log files on a CD or DVD, making it impossible to write new data into the same directory.

Another modification that would possibly make this script easier to use is to add a GUI with selection buttons for various functionality, such as where to store the output and report.

## Master Craftsman

### Parsing Event Logs

A project to parse not only Windows 2000, XP, and 2003 Event Logs, but also Windows Vista/2008 Event Logs in the same script could be accomplished by combining evt2xls.pl with code that Andreas Schuster has made available via his blog. In fact, knowing the "magic number" for Windows 2000, XP, and 2003 Event Logs would allow you to start by "looking at" the Event Log file in binary mode, and if you don't find that "magic number", then you could assume that the file is from a Windows Vista or 2008 system, and switch over to appropriate code.

# Parsing RAM Dumps

During incident response activities, the responder may opt to dump the contents of physical memory, or RAM, from a Windows system. This is done to preserve the contents of physical memory for later use and examination, and as stated by Aaron Walters and Nick Petroni during their Black Hat DC 2007[6] presentation, to answer new questions later. In some cases, the examiner has run strings.exe against the resulting file to attempt to locate passwords or other unique strings, or used regular expressions (regex's) to locate IP addresses, email addresses, etc. However, these simple searches constitute only the most rudimentary activities that can be performed when analyzing memory dumps. For example, we can extract a list of active processes from the memory dump, including the process memory and the executable image file for each (this is extremely useful when performing dynamic malware analysis).

---

[6] http://www.blackhat.com/html/bh-media-archives/bh-archives-2007.html#dc

# Lsproc.pl

Lsproc.pl is a Perl script I wrote in order to parse through a RAM dump from a Windows 2000 system and locate the remnants of processes that were running, or had exited, on the live system. Lsproc.pl is based in part on the original ptfinder.pl script written by Andreas Schuster and posted on his blog.[7]

```perl
#! c:\perl\bin\perl.exe
#---------------------------------------------------------------
# lsproc.pl - parse Windows 2000 phys. memory/RAM dump,
#      looking for processes.
#
# Version 0.1_2K 20060524
# Usage: lsproc <path_to_dump_file>
#
# copyright 2007 H. Carvey, keydet89@yahoo.com
#---------------------------------------------------------------
use strict;
print "lsproc - list processes from a Win2K dd-style RAM Dump (v.0.1_2K 20060524)\n";
print "Ex: lsproc <path_to_dump_file>\n";
print "\n";
my $file = shift || die "You must enter a filename.\n";
die "$file not found.\n" unless (-e $file);

my $record;
open(FH,"<",$file) || die "Could not open $file : $!\n";
binmode(FH);

my $offset = 0;

printf "%-4s %-6s %-6s %-20s %-10s %-20s\n","Type","PPID","PID","Name","Offset",
"Creation Time";
printf "%-4s %-6s %-6s %-20s %-10s %-20s\n","-" x 4,"-" x 4,"-" x 3,"-" x 4,"-"
x 6,"-" x 13;
while (! eof(FH)) {
  seek(FH,$offset,0);
  read(FH,$record,4);
  my ($type,$size) = unpack("CxCx",$record);
  if ($size == 0x1b && $type == 0x03) {
#      my $hdr = unpack("V",$record);
#      if ($hdr == 0x001b0003) {
# Possible EPROCESS block located, let's run some checks
```

---

[7] http://computer.forensikblog.de/en/topics/windows/memory_analysis/

```
#       printf "Possible EPROCESS block located at offset 0x%08x\n",$offset;
    my $data;
    seek(FH,$offset,0);
    my $bytes = read(FH,$data,0x290);
    if (0x290 == $bytes) {
      if (my %proc = isProcess($data)) {
        my $name;
        my $proctime;
        ($proc{createtime} == 0) ? ($proctime = "") : ($proctime = gmtime($proc
createtime}));
        ($proc{exitprocesscalled} == 1) ? ($name = $proc{name}."(x)") : ($name =
$proc{name});
        printf "%-4s %-6d %-6d %-20s 0x%08x %-20s\n","Proc",$proc{ppid},$proc{pid},
$name,$offset, $proctime;
        $offset += 0x290;
      }
      else {
        $offset += 4;
      }
    }
    else {
#       print "Too few bytes read.\n";
#       exit 1;
    }
  }
  elsif ($type == 0x06 && $size == 0x6c) {
# elsif ($hdr == 0x006c0006) {
# Possible ETHREAD found
    my $data;
    seek(FH,$offset,0);
    my $bytes = read(FH,$data,0x244);
    if ($bytes == 0x244) {
      if (my %thread = isThread($data)) {
        $offset += 0x244;
      }
      else {
        $offset += 4;
      }
    }
  }
  else {
```

```perl
# Increment the offset count by 4 bytes, or one DWORD
    $offset += 4;
  }
}
close(FH);
#----------------------------------------------------------------
# isProcess()
# check to see if we have a valid process (Win2K SP4)
# Input : 652 bytes starting at the offset
# Output: Hash containing EPROCESS block info, undef if not a valid
#         EPROCESS block
#----------------------------------------------------------------
sub isProcess {
  my $data = shift;
  my %proc = ();
  my $event1 = unpack("V",substr($data,0x13c,4));
  my $event2 = unpack("V",substr($data,0x164,4));
  if ($event1 == 0x40001 && $event2 == 0x40001) {
# Use this area to populate the EPROCESS structure
        my $name = substr($data,0x1fc,16);
        $name = ~ s/\00//g;
        $proc{name} = $name;
#       $proc{exitstatus} = unpack("V", substr($data,0x06c,4));
# Get Active Process Links for EPROCESS block
#       ($proc{flink},$proc{blink}) = unpack("VV",substr($data,0x0a0,8));
   my (@createTime)    = unpack("VV", substr($data,0x088,8));
   $proc{createtime}    = getTime($createTime[0],$createTime[1]);
# my (@exitTime)       = unpack("VV", substr($data,0x090,8));
# $proc{exittime}      = getTime($exitTime[0],$exitTime[1]);
# $proc{pObjTable}     = unpack("V",substr($data,0x128,4));
# $proc{pSectionHandle}= unpack("V",substr($data,0x1ac,4));
# $proc{pSecBaseAddr}  = unpack("V",substr($data,0x1b4,4));
   $proc{pid} = unpack("V",substr($data,0x09c,4));
   $proc{ppid} = unpack("V",substr($data,0x1c8,4));
#     ($proc{subsysmin},$proc{subsysmaj}) = unpack("CC",substr($data,0x212,2));
#       $proc{directorytablebase} = unpack("V",substr($data,0x018,4));
#       $proc{peb} = unpack("V",substr($data,0x1b0,4));
   $proc{exitprocesscalled} = unpack("C",substr($data,0x1aa,1));
   $proc{pimagefilename} = unpack("V",substr($data,0x284,4));
  }
```

```perl
    else {
# Not an EPROCESS block
  }
  return %proc;
}
#--------------------------------------------------------------
# isThread()
# check to see if we have a valid thread (Win2K SP4)
# Input : 0x244 bytes starting at the offset
# Output: Hash containing ETHREAD block info, undef if not a valid
#       ETHREAD block
#--------------------------------------------------------------
sub isThread {
      my $data = shift;
      my %thread = ();
#      my $ktimer = unpack("V",substr($data,0x0e8,4));
      my $sync1 = unpack("V",substr($data,0x190,4));
      my $sync2 = unpack("V",substr($data,0x1e8,4));

      if ($sync1 == 0x50005 && $sync2 == 0x50005) {
  ($thread{pid},$thread{tid}) = unpack("VV",substr($data,0x1e0,8));
  $thread{hasterminated} = unpack("V",substr($data,0x224,4));
  my (@createTime) = unpack("VV", substr($data,0x1b0,8));
  $thread{createtime} = getTime($createTime[0],$createTime[1]);
  my (@exitTime) = unpack("VV", substr($data,0x1b8,8));
  $thread{exittime} = getTime($exitTime[0],$exitTime[1]);
  $thread{hidefromdebugger} = unpack("C",substr($data,0x223,1));
  }
  return %thread;
}
#--------------------------------------------------------------
# getOffset()
# Get physical offset within dump, based on logical addresses
# Translates a logical address to a physical offset w/in the dump
# file
# Input : two addresses (ex: PEB and DirectoryTableBase)
# Output: offset within file
#--------------------------------------------------------------
sub getOffset {
  my $peb = shift;
  my $dtb = shift;
```

```perl
  my $pdi = $peb >> 22 & 0x3ff;
  my $pda = $dtb + ($pdi * 4);
  seek(FH,$pda,0);
  read(FH,$record,4);
  my $pde = unpack("V",$record);
# Determine page size if needed
# $pde & 0x080; if 1, page is 4Mb; else, 4Kb
# Check to see if page is present
    if ($pde & 0x1) {
            my $pti = $peb >> 12 & 0x3ff;
            my $ptb = $pde >> 12;
            seek(FH,($ptb * 0x1000) + ($pti * 4),0);
            read(FH,$record,4);
            my $pte = unpack("V",$record);
            if ($pte & 0x1) {
                my $pg_ofs = $peb & 0x0fff;
                return ((($pte >> 12) * 0x1000) + $pg_ofs);
    }
    else {
            return 0;
    }
  }
  else {
            return 0;
  }
}
#------------------------------------------------------------
# getTime()
# Get Unix-style date/time from FILETIME object
# Input : 8 byte FILETIME object
# Output: Unix-style date/time
#------------------------------------------------------------
sub getTime() {
  my $lo = shift;
  my $hi = shift;
  my $t;
  if ($lo == 0 && $hi == 0) {
    $t = 0;
```

```
  } else {
    $lo -= 0xd53e8000;
    $hi -= 0x019db1de;
    $t = int($hi*429.4967296 + $lo/1e7);
  };
  $t = 0 if ($t < 0);
  return $t;
}

#------------------------------------------------------------
# _uniToAscii()
# Input : Unicode string
# Output: ASCII string
# Removes every other \00 from Unicode strings, returns ASCII string
#------------------------------------------------------------
sub _uniToAscii {
  my $str = $_[0];
  my $len = length($str);
  my $newlen = $len - 1;
  my @str2;
  my @str1 = split(//,$str,$len);
  foreach my $i (0..($len - 1)) {
    if ($i % 2) {
# In a Unicode string, the odd-numbered elements of the list will be \00
# so just drop them
    }
    else {
      push(@str2,$str1[$i]);
    }
  }
  return join('',@str2);
}
```

In order to run the lsproc.pl script, launch it from the command line and pass it a single argument, that being the path to the RAM dump file:

```
C:\Perl\forensics>lsproc.pl d:\cases\case007\ramdump.img
```

When run against the first RAM dump available from the DFRWS 2005 Memory Challenge web site[8] (renamed above to 'ramdump.img'), lsproc.pl produces output similar to the following (excerpted for the sake of brevity):

---

[8]  http://www.dfrws.org/2005/challenge/

| Proc | 228 | 672 | WinMgmt.exe | 0x0017dd60 | Sun | Jun | 5 | 00:32:59 | 2005 |
|------|-----|------|-------------|------------|-----|-----|---|----------|------|
| Proc | 820 | 324 | helix.exe | 0x00306020 | Sun | Jun | 5 | 14:09:27 | 2005 |
| Proc | 0 | 0 | Idle | 0x0046d160 | | | | | |
| Proc | 600 | 668 | UMGR32.EXE | 0x0095f020 | Sun | Jun | 5 | 00:55:08 | 2005 |
| Proc | 324 | 1112 | cmd2k.exe | 0x00dcc020 | Sun | Jun | 5 | 14:14:25 | 2005 |
| Proc | 156 | 176 | winlogon.exe | 0x01045d60 | Sun | Jun | 5 | 00:32:44 | 2005 |
| Proc | 144 | 164 | winlogon.exe | 0x0104ca00 | Fri | Jun | 3 | 01:25:54 | 2005 |
| Proc | 156 | 180 | csrss.exe | 0x01286480 | Sun | Jun | 5 | 00:32:43 | 2005 |
| Proc | 8 | 156 | smss.exe | 0x012b62c0 | Sun | Jun | 5 | 00:32:40 | 2005 |
| Proc | 0 | 8 | System | 0x0141dc60 | | | | | |
| Proc | 1112 | 1152 | dd.exe(x) | 0x019d1980 | Sun | Jun | 5 | 14:14:38 | 2005 |
| Proc | 228 | 592 | dfrws2005.exe | 0x02138640 | Sun | Jun | 5 | 01:00:53 | 2005 |
| Proc | 820 | 1076 | cmd.exe | 0x02138c40 | Sun | Jun | 5 | 00:35:18 | 2005 |

The first column of the output is the object that was found; in this case, only the processes are shown (I removed the display of threads as the output got too verbose and difficult to understand). The second column shows the parent process identifier (PID) while the third column shows the PID of the process. The fourth column shows the name of the process, and the fifth column displays the offset of where the process was located within the RAM dump file. Finally, the sixth column displays the creation time of the process (extracted from the process structure, or "block") in UTC format. You'll notice that one of the displayed processes, specifically "dd.exe", has an "(x)" after the name. This indicates that the process was exited by the time the RAM dump was made, and that the process will have an exited time associated with it.

> **NOTE**
>
> Lsproc.pl and the other associated scripts for parsing information from RAM dumps were designed to work on RAM dumps from Windows 2000 systems only. There are significant changes between operating system versions with regards to how the operating system manages memory – not just between versions, but in some instances, between Service Packs! Some additional work is required to get these scripts to work on systems other than Windows 2000.

# Lspi.pl

The name of the lspi.pl Perl script stands for "list process image"; not the most descriptive title, particularly as only four letters, I know, but I had to come up with something! The really cool thing is that aside from the name, what the script does is even more cool…it extracts, if possible, the actual image file (i.e., *.exe file) for a once-running process from a RAM dump.

```perl
#! c:\perl\bin\perl.exe
#-----------------------------------------------------------------------------
# lspi.pl - parse process image from a Windows 2000 phys. memory/RAM dump,
#      (LiSt Process Image)
#
# Version 0.4
#
# Usage: lspi.pl <filename> <offset>
#      Determine the offset of the the process you're interested in by
#      running lsproc.pl first
#
# Changelog:
# 20060721 - created
#
# copyright 2007 H. Carvey, keydet89@yahoo.com
#-----------------------------------------------------------------------------
use strict;
print "lspi - list Windows 2000 process image (v.0.4 - 20060721)\n";
print "Ex: lspi <path_to_dump_file> <offset_from_lsproc>\n";
print "\n";
my $file = shift || die "You must enter a filename.\n";
die "$file not found.\n" unless (-e $file);
my $offset = hex(shift) || die "You must enter a process offset.\n";
my $data;
my $error;
my ($size,$type);

#-----------------------------------------------------------------------------
# Global Variables
#-----------------------------------------------------------------------------
my $pagecount = 0;
my %imagepages = ();
my @pagedout = ();
my $outfile;

open(FH,"<",$file) || die "Could not open $file : $!\n";
binmode(FH);
seek(FH,$offset,0);
```

```perl
read(FH,$data,4);
my ($type,$size) = unpack("CxCx",$data);
if ($size == 0x1b && $type == 0x03) {
  seek(FH,$offset,0);
  my $bytes = read(FH,$data,0x290);
  if (0x290 == $bytes) {
    if (my %proc = isProcess($data)) {
      print "Process Name : ".$proc{name}."\n";
      $outfile = $proc{name}."\.img";
      print "PID      : ".$proc{pid}."\n";
      my $dtb = $proc{directorytablebase};
      printf "DTB      : 0x%08x\n",$dtb;
      my $peb_ofs = getOffset($proc{peb},$dtb);
      die "The page located at the address for the PEB has been paged out.\n"
if ($peb_ofs == 0);
      printf "PEB      : 0x%08x (0x%08x)\n",$proc{peb},$peb_ofs;
# Get specific info from the PEB
      if ($peb_ofs != 0x0) {
        my %peb_data = getPEBData($peb_ofs,$dtb);
        my $imgbaseofs = getOffset($peb_data{img_base_addr},$dtb);
        die "The page located at the ImageBaseAddress for this process has been
paged out.\n" if ($imgbaseofs == 0);
        printf "ImgBaseAddr : 0x%08x (0x%08x)\n",$peb_data{img_base_addr},
$imgbaseofs;
        print "\n";
        if ($imgbaseofs != 0x00 && getImgBase($imgbaseofs)) {
# We're now ready to begin processing
          $pagecount++;
          $imagepages{$pagecount} = $imgbaseofs;
# Read in the first 4K page located at the ImageBaseAddress offset
          seek(FH,$imgbaseofs,0);
          read(FH,$data,0x1000);
# Check the NT Header
          my $e_lfanew = unpack("V",substr($data,0x3c,4));
          printf "e_lfanew = 0x%x\n",$e_lfanew;
          my $nt = unpack("V",substr($data,$e_lfanew,4));
          die "Not an NT header.\n" if ($nt != 0x4550);
          printf "NT Header = 0x%x\n",$nt;
          print "\n";
          print "Reading the Image File Header\n";
          my %ifh;
```

```
($ifh{machine},$ifh{number_sections},$ifh{datetimestamp},$ifh{ptr_symbol_table},
 $ifh{number_symbols},$ifh{size_opt_header},$ifh{characteristics})
        = unpack("vvVVVvv",substr($data,$e_lfanew + 4,20));
     print "Sections = $ifh{number_sections}\n";
     printf "Opt Header Size = 0x%08x (".$ifh{size_opt_header}." bytes)
\n",$ifh{size_opt_header};
     print "Characteristics: \n";
# Translate the image file header characteristics
     my @char = getFileHeaderCharacteristics($ifh{characteristics});
     foreach (@char) {print "\t$_\n";}
     print "\n";
     print "Machine = ".getFileHeaderMachine($ifh{machine})."\n";
     print "\n";
     print "Reading the Image Optional Header\n";
     print "\n";
     my $opt_hdr = unpack("v",substr($data, $e_lfanew + 24,2));
     printf "Opt Header Magic = 0x%x\n",$opt_hdr;
     my %opt32 = ();
 ($opt32{magic},$opt32{majlinkver},$opt32{minlinkver},$opt32{codesize},
 $opt32{initdatasz},$opt32{uninitdatasz},$opt32{addr_entrypt},$opt32{codebase},
 $opt32{database},$opt32{imagebase},$opt32{sectalign},$opt32{filealign},
 $opt32{os_maj},$opt32{os_min},$opt32{image_maj},$opt32{image_min},
 $opt32{image_sz},$opt32{head_sz},$opt32{checksum},$opt32{subsystem},
 $opt32{dll_char},$opt32{rva_num}) = unpack("vCCV9v4x8V3vvx20Vx4",substr($data,
$e_lfanew + 24,$ifh{size_opt_header}));
     print "Subsystem     : ".getOptionalHeaderSubsystem($opt32{subsystem})."\n";
     printf "Entry Pt Addr : 0x%08x\n",$opt32{addr_entrypt};
     printf "Image Base    : 0x%08x\n",$opt32{imagebase};
     printf "File Align    : 0x%08x\n",$opt32{filealign};
# get Data Directories
    print "\n";
     print "Reading the Image Data Directory information\n";
     my %dd = ();
     my @dd_names = qw/ExportTable ImportTable ResourceTable ExceptionTable
        CertificateTable BaseRelocTable DebugTable ArchSpecific
        GlobalPtrReg TLSTable LoadConfigTable BoundImportTable
        IAT DelayImportDesc CLIHeader unused/;
     my @rva_list = unpack("VV" x $opt32{rva_num},substr($data,$e_lfanew + 24 +
96,8*$opt32{rva_num}));
     foreach my $i (0..($opt32{rva_num} - 1)) {
       $dd{$dd_names[$i]}{rva} = $rva_list[($i*2)];
```

```
            $dd{$dd_names[$i]}{size} = $rva_list[($i*2)+1];
        }
        print "\n";
        printf "%-20s %-10s %-10s\n","Data Directory","RVA","Size";
        printf "%-20s %-10s %-10s\n","-" x 14,"-" x 3, "-" x 4;
        foreach my $name (keys %dd) {
            printf "%-20s 0x%08x 0x%08x\n",$name,$dd{$name}{rva},$dd{$name}{size};
        }
# Read section headers
        print "\n";
        print "Reading Image Section Header information\n";
        print "\n";
        my $num = $ifh{number_sections};
        my $size = 40;
        my $ofs = $e_lfanew + 24 + 96 + 8*$opt32{rva_num};
        my $sect = substr($data,$ofs,$num * $size);
        my %sections = getImageSectionHeaders($sect,$num);

        printf "%-8s %-10s %-10s %-10s %-10s %-10s\n","Name","Virt Sz","Virt
Addr","rData Ofs","rData Sz","Char";
        printf "%-8s %-10s %-10s %-10s %-10s %-10s\n","-" x 4,"-" x 7,"-" x 9,
"-" x 9,"-" x 8,"-" x 4;
        my %sec_order = ();
        foreach my $sec (keys %sections) {
            printf "%-8s 0x%08x 0x%08x 0x%08x 0x%08x 0x%08x\n",$sec,$sections{$sec}
{virt_sz},
            $sections{$sec}{virt_addr},$sections{$sec}{rdata_ptr},$sections{$sec}
{rdata_sz},
            $sections{$sec}{characteristics};
            $sec_order{$sections{$sec}{virt_addr}} = $sec;
        }
        print "\n";
# Now that we have information from the section headers, we need calculate the
offsets of the pages
# within the dump file, and check to see if any of the pages have been paged out.
        foreach my $order (sort {$a <=> $b} keys %sec_order) {
            my $sec = $sec_order{$order};
            my $num_pages = $sections{$sec}{rdata_sz} / 0x1000;
            foreach my $n (0..($num_pages - 1)) {
                my $page = $peb_data{img_base_addr} + $sections{$sec}{virt_addr} +
(0x1000 * $n);
                my $offset = getOffset($page, $dtb);
                if ($offset == 0) {
```

```perl
    push(@pagedout,$page);
            }
            else {
    $pagecount++;
    $imagepages{$pagecount} = $offset;
            }
#      seek(FH,$offset,0);
#      read(FH,$data,0x1000);
#      syswrite(OUT,$data,length($data));
          }
        }
        if (scalar(@pagedout) > 0) {
          print "There are ".scalar(@pagedout)." pages paged out of physical
memory.\n";
          map{printf "\t0x%08x\n",$_}(@pagedout);
          print "If any pages are paged out, the image file cannot be completely
reassembled.\n";
        }
        else {
          print "Reassembling image file into $outfile\n";
          open(OUT,">",$outfile) || die "Could not open $outfile: $!\n";
          binmode(OUT);
          my $size = 0;
          foreach my $i (sort {$a <=> $b} keys %imagepages) {
    seek(FH,$imagepages{$i},0);
    read(FH,$data,0x1000);
    syswrite(OUT,$data,length($data));
            $size += length($data);
          }
          close(OUT);
          print "Bytes written = $size\n";
          print "New file size = ".(stat($outfile))[7]."\n";
        }
      }
    }
  }
}
close(FH);
#-------------------------------------------------------------------
# isProcess()
```

```perl
# check to see if we have a valid process (Win2K SP4)
# Input : 652 bytes starting at the offset
# Output: Hash containing EPROCESS block info, undef if not a valid
#        EPROCESS block
#-------------------------------------------------------------------
sub isProcess {
  my $data = shift;
  my %proc = ();
  my $event1 = unpack("V",substr($data,0x13c,4));
  my $event2 = unpack("V",substr($data,0x164,4));

  if ($event1 == 0x40001 && $event2 == 0x40001) {
# Use this area to populate the EPROCESS structure
    my $name = substr($data,0x1fc,16);
    $name =~ s/\00//g;
    $proc{name} = $name;
#    $proc{exitstatus} = unpack("V", substr($data,0x06c,4));
# Get Active Process Links for EPROCESS block
#       ($proc{flink},$proc{blink}) = unpack("VV",substr($data,0x0a0,8));
    $proc{pid} = unpack("V",substr($data,0x09c,4));
#        $proc{ppid} = unpack("V",substr($data,0x1c8,4));
#        ($proc{subsysmin},$proc{subsysmaj}) = unpack("CC",substr($data,0x212,2));
    $proc{directorytablebase} = unpack("V",substr($data,0x018,4));
  $proc{peb} = unpack("V",substr($data,0x1b0,4));
  $proc{exitprocesscalled} = unpack("C",substr($data,0x1aa,1));
#       $proc{pimagefilename} = unpack("V",substr($data,0x284,4));
  }
  else {
# Not an EPROCESS block
  }
  return %proc;
}
#-------------------------------------------------------------------
# getOffset()
# Get physical offset within dump, based on logical addresses
# Translates a logical address to a physical offset w/in the dump
#      file
# Input : two addresses (ex: PEB and DirectoryTableBase)
# Output: offset within file
#-------------------------------------------------------------------
sub getOffset {
```

```perl
  my $peb = shift;
  my $dtb = shift;
  my $record;
  my $pdi = $peb >> 22 & 0x3ff;
  my $pda = $dtb + ($pdi * 4);
  seek(FH,$pda,0);
  read(FH,$record,4);
  my $pde = unpack("V",$record);
# Determine page size if needed
# $pde & 0x080; if 1, page is 4Mb; else, 4Kb
# Check to see if page is present
  if ($pde & 0x1) {
    my $pti = $peb >> 12 & 0x3ff;
    my $ptb = $pde >> 12;
    seek(FH,($ptb * 0x1000) + ($pti * 4),0);
    read(FH,$record,4);
    my $pte = unpack("V",$record);
    if ($pte & 0x1) {
      my $pg_ofs = $peb & 0x0fff;
      return ((($pte >> 12) * 0x1000) + $pg_ofs);
    }
    else {
      return 0;
    }
  }
  else {
    return 0;
  }
}
#----------------------------------------------------------------
# getPEBData()
# Input : physical offset to PEB
# Output: data from PEB (Note: Virtual addresses are not
#       translated to physical offsets in this subroutine)
#----------------------------------------------------------------
sub getPEBData() {
  my $ofs = shift;
  my $dtb = shift;
  my %peb = ();
  seek(FH,$ofs,0);
```

```perl
  my $record;
  read(FH,$record,20);
   ($peb{inheritedaddrspace},$peb{readimgfileexecopts},$peb{beingdebugged},
$peb{mutant},
    $peb{img_base_addr},$peb{peb_ldr},$peb{params}) = unpack("C3xV4",$record);
  return %peb;
}
#------------------------------------------------------------------
# getImgBase()
# Read 4K at image base offset (from PEB)
# Input : Physical offset to the image base addr
# Output: dump of memory
#------------------------------------------------------------------
sub getImgBase {
  my $ofs = shift;
  my $data;
  seek(FH,$ofs,0);
  read(FH,$data,2);
  my $mz = unpack("v",$data);
  if ($mz == 0x00005a4d) {
    return 1;
  }
  else {
    return 0;
  }
}
#------------------------------------------------------------------
sub getFileHeaderCharacteristics {
  my $char = shift;
  my @list = ();
  my %chars = (0x0001 => "IMAGE_FILE_RELOCS_STRIPPED",
               0x0002 => "IMAGE_FILE_EXECUTABLE_IMAGE",
               0x0004 => "IMAGE_FILE_LINE_NUMS_STRIPPED",
               0x0008 => "IMAGE_FILE_LOCAL_SYMS_STRIPPED",
               0x0010 => "IMAGE_FILE_AGGRESIVE_WS_TRIM",
               0x0020 => "IMAGE_FILE_LARGE_ADDRESS_AWARE",
               0x0080 => "IMAGE_FILE_BYTES_REVERSED_LO",
               0x0100 => "IMAGE_FILE_32BIT_MACHINE",
               0x0200 => "IMAGE_FILE_DEBUG_STRIPPED",
               0x0400 => "IMAGE_FILE_REMOVABLE_RUN_FROM_SWAP",
```

```perl
          0x0800 => "IMAGE_FILE_NET_RUN_FROM_SWAP",
          0x1000 => "IMAGE_FILE_SYSTEM",
          0x2000 => "IMAGE_FILE_DLL",
          0x4000 => "IMAGE_FILE_UP_SYSTEM_ONLY",
          0x8000 => "IMAGE_FILE_BYTES_REVERSED_HI");
  foreach my $c (keys %chars) {
    push(@list,$chars{$c}) if ($char & $c);
  }
  return @list;
}
sub getFileHeaderMachine {
  my $word = shift;
  my %mach = (0x014c  => "IMAGE_FILE_MACHINE_I386",
              0x014d  => "IMAGE_FILE_MACHINE_I860",
              0x0184  => "IMAGE_FILE_MACHINE_ALPHA",
              0x01c0  => "IMAGE_FILE_MACHINE_ARM",
              0x01c2  => "IMAGE_FILE_MACHINE_THUMB",
              0x01f0  => "IMAGE_FILE_MACHINE_POWERPC",
              0x0284  => "IMAGE_FILE_MACHINE_ALPHA64",
              0x0200  => "IMAGE_FILE_MACHINE_IA64",
              0x8664  => "IMAGE_FILE_MACHINE_AMD64");
  foreach my $m (keys %mach) {
    return $mach{$m} if ($word & $m);
  }
}
sub getOptionalHeaderSubsystem {
  my $word = shift;
  my %subs = (0 => "IMAGE_SUBSYSTEM_UNKNOWN",
              1 => "IMAGE_SUBSYSTEM_NATIVE",
              3 => "IMAGE_SUBSYSTEM_WINDOWS_CUI",
              2 => "IMAGE_SUBSYSTEM_WINDOWS_GUI",
              5 => "IMAGE_SUBSYSTEM_OS2_CUI",
              7 => "IMAGE_SUBSYSTEM_POSIX_CUI",
              8 => "IMAGE_SUBSYSTEM_NATIVE_WINDOWS",
              9 => "IMAGE_SUBSYSTEM_WINDOWS_CE_GUI",
              14 => "IMAGE_SUBSYSTEM_XBOX");

  foreach my $s (keys %subs) {
    return $subs{$s} if ($word == $s);
  }
}
```

```perl
sub getImageSectionHeaders {
  my $data    = shift;
  my $num     = shift;
# Each section is 40 bytes in size, and all sections are contiguous
  my $sec_sz  = 40;
  my %sec  = ();
  foreach my $i (0..($num - 1)) {
    my ($name,$virt_sz,$virt_addr,$rdata_sz,$rdata_ptr,$char)
      = unpack("a8V4x12V",substr($data,$i * $sec_sz,$sec_sz));
    $name =~ s/\00+$//;
    $sec{$name}{virt_sz}        = $virt_sz;
    $sec{$name}{virt_addr}      = $virt_addr;
    $sec{$name}{rdata_sz}       = $rdata_sz;
    $sec{$name}{rdata_ptr}      = $rdata_ptr;
    $sec{$name}{characteristics} = $char;
  }
  return %sec;
}
```

Lspi.pl is a bit more complex that many of the other Perl scripts I've written, however, it's fairly easy to run. To launch the script, all you need to provide is the path to the RAM dump file, and the offset of where the process was located within the RAM dump file. This second argument can be populated from the output of the lsproc.pl Perl script; therefore, in order to run lspi.pl, you first have to run lsproc.pl.

Using the same RAM dump file (retrieved from the DFRWS 2005 Memory Challenge site and renamed to ramdump.img), you can extract the executable image for the running dd.exe process that was actually used to create the RAM dump itself, using the following command:

```
C:\Perl\forensics>lspi.pl d:\case007\ramdump.img 0x0414dd60
```

Or, you can extract the executable image file for the nc.exe process:

```
C:\Perl\forensics>lspi.pl d:\case007\ramdump.img 0x0625d3c0
```

Again, in order to get the hex values for the second argument in each of the above commands, the examiner would need to run lsproc.pl first, and then use the output of that script to provide the necessary values for lspi.pl.

Now, what makes lspi.pl so complex from a coding perspective is that the script does some binary file data parsing of its own, beyond simply accessing the RAM dump file. The script also has to parse the portable executable (PE) file header (the PE file header is defined by Microsoft) and reassemble the various bits and pieces of the executable image file based on the "map" provided by the header information. The actual segments of the executable image file are maintained in memory "pages", sections that are each 4096 bytes (4Kb) in size. If all of the necessary memory pages are not available in the RAM dump, such as if the memory manager swapped those pages out to the pagefile, lspi.pl will not be able to reassemble the complete executable image file and will quit with a message to that effect.

These two Perl scripts (lsproc.pl and lspi.pl) are available, along with several other Perl scripts that can be used to parse data from RAM dumps, on the SourceForge site for the WindowsIR[9] project.

## Master Craftsman

### Extracting other data from RAM

RAM dumps are very often full of much more than just processes and network connections. Opening a RAM dump in your favorite hex editor will reveal what might appear to be Event Log event records, as well as Registry keys. In much the same way as processes were located in a RAM dump, these other objects can be located and parsed, as well. This can be very useful in an investigation as these objects contain timestamp information and may be used to establish or correlate a timeline of activity on the system.

# ProScripts

As mentioned in Part I, ProScripts are essentially Perl scripts that provide a scripting capability for the ProDiscover forensic analysis application provided by Technology Pathways. The ProDiscover forensic analysis application can be extremely useful, and the addition of the Perl scripting language allows the examiner to leverage the power of Perl in conjunction with the ProDiscover product. This gives the examiner the ability to extend the ProDiscover application to incredible levels of usability.

Graphical user interface (GUI) programming is beyond the scope of this book, but I have tested the use of a GUI in conjunction with ProDiscover ProScripts

---

[9] http://sourceforge.net/project/showfiles.php?group_id=164158

(using the Win32::GUI module) and they have worked very well together. Taking this to greater lengths would allow the examiner to extend the usability of ProDiscover even further, all thanks to the power of Perl!

# Uassist.pl

In Part I, as well as previously in this Part, I provided Perl scripts that could be used to extract information from the UserAssist keys in the Registry. In Part I, the uassist. pl Perl script could be run on a live system to retrieve the contents of the UserAssist key for the currently logged on user, and earlier in this Part, the uassist.pl Perl script used the Parse::Win32Registry module to retrieve the same information from an arbitrary NTUSER.DAT file that had been extracted from an acquired image. For the sake of completeness, I wanted to provide a ProScript that you could use with ProDiscover and parse the UserAssist keys from the user Registry files:

```
#! c:\Perl\bin\perl.exe
#------------------------------------------------------------------
# UAssist.pl, version 0.11
#
# Copyright 2006-2007 H. Carvey, keydet89@yahoo.com
#------------------------------------------------------------------
use ProScript;
PSDisplayText("UserAssist.pl v.0.11_20060522");
PSDisplayText("ProScript to parse the UserAssist keys within each
user's Registry file");
PSDisplayText("decrypt the values, and display the data as a GMT time,
where applicable.");
PSDisplayText("Also, values with time-stamped data are sorted by time,
in reverse order, so");
PSDisplayText("that timelining the user activity is done more readily.");
PSDisplayText("\n");
my @sids = ();
$numRegs = PSGetNumRegistries();
if ($numRegs == 0) {
  PSDisplayText("No registries to process");
  return;
}
$regName = PSGetRegistryAt(0);
PSRefreshRegistry($regName);
#------------------------------------------------------------------
my $hiveName = "HKEY_Users";
```

```perl
my $rHandle = PSOpenRegistry($regName, $hiveName);
my @sids = ();

if ($rHandle == 0) {
  PSDisplayText("Unable to locate registry key");
  return;
}
else {
  PSDisplayText("Registry opened succesfully.");
}
#Successfully opened the key. Now, enumerate the key.
while (1) {
  $RegKeyInfo = &ProScript::PSReadRegistry($rHandle);
  last if ($RegKeyInfo->{nType} == -1);
  push(@sids,$RegKeyInfo->{strRegName}) if (length($RegKeyInfo->
{strRegName}) > 20);
}
PSCloseHandle($rHandle);
PSDisplayText("Registry handle closed.");

# Now that we have the SIDs, let's enumerate through the keys
my @guids = ("{5E6AB780-7743-11CF-A12B-00AA004AE837}\\Count",
     "{75048700-EF1F-11D0-9888-006097DEACF9}\\Count");
my $key_path = "\\Software\\Microsoft\\Windows\\CurrentVersion\\Explorer\\
UserAssist\\";
foreach my $sid (@sids) {

# use %sorter as a hash-of-arrays data structure for maintaining a
sorted list of
# times
  my %sorter = ();

  foreach my $g (@guids) {
    my $key = $hiveName."\\".$sid.$key_path.$g;
    PSDisplayText("Key : $key");
    my $rHandle = PSOpenRegistry($regName,$key);
    while (1) {
  $RegKeyInfo = &ProScript::PSReadRegistry($rHandle);
  last if ($RegKeyInfo->{nType} == -1);
  next if ($RegKeyInfo->{strRegName} eq "(Default)");
  my $value = $RegKeyInfo->{strRegName};
  $value =~ tr/N-ZA-Mn-za-m/A-Za-z/;
#      PSDisplayText("\t".$value);
  my $data = $RegKeyInfo->{strValueData};
  my $l = length($data);
```

```perl
     if ($l == 16) {
        my @vals = unpack("V4",substr($data,0,16));
      my $gtime = _getTimeDate($vals[3],$vals[2]);
       if ($gtime > 0) {
         PSDisplayText("\t".$value." --> ".gmtime($gtime));
# The following code adds the ROT-13 (decrypted) entry to an array in the
hash-of-arrays
# data structure
        if ($g eq "{75048700-EF1F-11D0-9888-006097DEACF9}\\Count") {
          push(@{$sorter{$gtime}},$value);
        }
       }
       else {
         PSDisplayText("\t".$value);
       }
     }
       }
     PSCloseHandle($rHandle);
    }
   PSDisplayText("\n");
# Display the time-based entries in reverse order, listing the entries that
# were accessed at that date/time beneath the time
   PSDisplayText("Time-sorted Entries");
   foreach my $item (reverse sort {$a <=> $b} keys %sorter) {
     PSDisplayText(" --> ".gmtime($item));
     foreach my $pdl (@{$sorter{$item}}) {
       PSDisplayText("\t --> $pdl");
     }
   }
   PSDisplayText("\n");
}
#---------------------------------------------------------------
# _getTimeDate()
# Input : 2 DWORDs, each containing half of the LastWrite time
# Output: readable GMT time string
#---------------------------------------------------------------
sub _getTimeDate {
# Borrowed from Andreas Schuster's ptfinder code
   my $Hi = shift;
   my $Lo = shift;
```

```
  my $t;
 if (($Lo == 0) and ($Hi == 0)) {
   $t = 0;
 }
 else {
   $Lo -= 0xd53e8000;
   $Hi -= 0x019db1de;
   $t = int($Hi*429.4967296 + $Lo/1e7);
 }
 $t = 0 if ($t < 0);
 return $t;
}
```

To run this script, simply open a project or a case in ProDiscover (the uassist.pl ProScript was written and tested on ProDiscover Incident Response Edition version 4.89), click the Run ProScript button on the button bar (illustrated in Figure II.3), and in the Run ProScript dialog, choose the path the uassist.pl ProScript (illustrated in Figure II.4).

**Figure II.3** ProDiscover "Run ProScript" Button

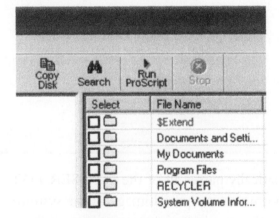

**Figure II.4** ProDiscover "Run ProScript" Dialog

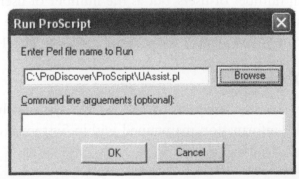

Once you hit the "OK" button, the script will run ... no arguments are required for this ProScript. The uassist.pl ProScript will parse each NTUSER.DAT file from each user profile within the ProDiscover project, and display all entries, followed by the reverse time-sorted entries, as shown in Figure II.5.

**Figure II.5** Output of uassist.pl ProScript

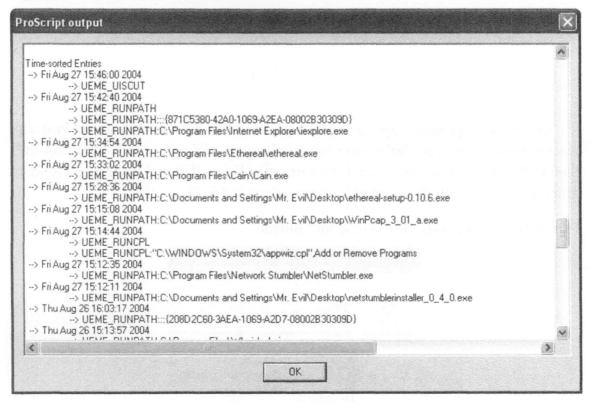

As uassist.pl automatically parses all of the NTUSER.DAT files for all of the local users on the system (within the acquired image of the system), there can be quite a bit of output displayed. This version of the script is great for examinations of systems with a relatively small number of users, but as the number of users increases, you may want to extract the specific NTUSER.DAT files from the image and parse them using the version of the uassist.pl Perl script listed earlier in this Part.

# SysRestore.pl

One of the aspects of forensic analysis of Windows XP systems that I've found to be extremely useful is the information maintained within System Restore Points.[10]

---

[10]  http://technet.microsoft.com/en-us/library/bb490854.aspx

System Restore Points allow the user to "back out" of a software installation and roll back the system state to a previous time (say, 3 days ago) when the system was known to be functioning properly. While this is extremely useful to the user, it can also be extremely useful to an examiner.

One of the more useful aspects of the restore points is that, as long as disk space is available, Windows XP will create a restore point every 24 hours, as well as when software is installed or uninstalled. Knowing this, the examiner can parse through each restore point, extracting information about the restore point from the rp.log file. The sysrestore.pl ProScript allows you to do this quickly and easily through the ProDiscover interface.

```perl
#! c:\perl\bin\perl.exe
#-----------------------------------------------------------
# SysRestore.pl, version 0.1_20061026
# ProScript to parse the System Restore subdirectories for
# rp.log files, and
# then parse the files for description and creation time info
#
# Copyright 2006-2007 H. Carvey, keydet89@yahoo.com
#-----------------------------------------------------------
use ProScript;
PSDisplayText("SysRestore.pl v. 0.1_20061026");
PSDisplayText("ProScript to parse through the System Restore
subdirectories on Windows XP");
PSDisplayText("systems and return the type, description and
creation time from each rp\.log files");
PSDisplayText("\n");
PSDisplayText("Restore Point Types:");
PSDisplayText("0 - Application Install");
PSDisplayText("1 - Application Uninstall");
PSDisplayText("7 - System CheckPoint");
PSDisplayText("10 - Device Driver Install");
PSDisplayText("12 - Modify Settings");
PSDisplayText("13 - Cancelled Operation");
PSDisplayText("\n");
#-----------------------------------------------------------
# Get the SystemRoot value
my %sysinfo = ();
$numRegs = PSGetNumRegistries();
if ($numRegs == 0) {
    PSDisplayText("No registries to process");
    return;
```

```perl
}
$regName = PSGetRegistryAt(0);
PSRefreshRegistry($regName);
my $keyName = "HKEY_LOCAL_MACHINE\\Software\\Microsoft\\Windows
NT\\CurrentVersion";
my $rHandle = PSOpenRegistry($regName, $keyName);

if ($rHandle == 0) {
  PSDisplayText("Unable to locate registry key");
  return;
}
else {
#       PSDisplayText("Registry opened succesfully.");
}

while (1) {
    $RegKeyInfo = &ProScript::PSReadRegistry($rHandle);
  last if ($RegKeyInfo->{nType} == -1);
  next if ($RegKeyInfo->{nType} == PS_TYPE_KEY);
  my $value = $RegKeyInfo->{strRegName};
  my $data = $RegKeyInfo->{strValueData};
# PSDisplayText($value." -> ".$data);
  $sysinfo{$value} = $data;
}
PSCloseHandle($rHandle);
#----------------------------------------------------------------
# Now we have a %sysinfo hash, and all we really want is the
# "SystemRoot" value
my $sysroot = $sysinfo{"SystemRoot"};
my $drive = (split(/:/,$sysinfo{"SystemRoot"},2))[0];
# $drive should now just be a drive letter

my $objectName = PSGetObjectName(0);
my $path = $objectName."\\".$drive.":\\System Volume Information";
#----------------------------------------------------------------
# First, we need to get the name of the _restore directory
#----------------------------------------------------------------
my $pHandle = PSOpenDir($path,0);
if ($pHandle == NULL) {
  PSDisplayText("$path not opened.");
}
my $rest = "_restore";
my $restoredir;
```

```perl
my $tag = 1;
while ($tag) {
  my $file = &ProScript::PSReadDirectory($pHandle);
  $tag = 0 if ($file == NULL || $file->{strName} eq "");
  $restoredir = $file->{strName} if ($file->{bIsDirectory} && $file->{strName} =~ m~
^$rest/i);
#        PSDisplayText("Name : $file->{strName}");
}
PSCloseHandle($pHandle);
$path = $path."\\".$restoredir."\\";

#----------------------------------------------------------------
# Now, we need to get the list of subdirectories
#----------------------------------------------------------------
my @rpdirs = ();
my $rpdir = "RP";
my $pHandle = PSOpenDir($path,0);
if ($pHandle == NULL) {
  PSDisplayText("$path not opened.");
}
my $tag = 1;
while ($tag) {
  my $file = &ProScript::PSReadDirectory($pHandle);
  $tag = 0 if ($file == NULL || $file->{strName} eq "");
  push(@rpdirs,$file->{strName}) if ($file->{bIsDirectory}
&& $file->{strName} =~ m/^$rpdir/);
#        PSDisplayText("Name : $file->{strName}");
}
PSCloseHandle($pHandle);
foreach my $rp (@rpdirs) {
  $rp_path = $path.$rp."\\rp\.log";
  my $type = getType($rp_path);
  my $descr = getRpDescr($rp_path);
  my $creation = getCreationTime($rp_path);
  PSDisplayText($rp." ".$type." ".$creation." (UTC) ".$descr);
}
#----------------------------------------------------------------
# getType()
# Read the rp.log file to get the restore point type
#----------------------------------------------------------------
```

```perl
sub getType {
  my $path = shift;
  my $type = 0;
  if (my $oFile = PSOpen($path)) {
    if (PSSeek($oFile,0x04,0,PS_FILE_BEGIN)) {
      my $buffer = PSReadRaw($oFile,4);
      PSCloseHandle($oFile);
      $type = unpack("V",$buffer);
    }
    else {
      PSDisplayText("File seek to first offset failed.");
    }
  }
  else {
    PSDisplayText("File could not be opened.");
  }
  return $type;
}
#-----------------------------------------------------------------
# getCreationTime()
# Read the rp.log file to get the description and creation
# date
#-----------------------------------------------------------------
sub getCreationTime {
  my $path = shift;
  my $t_val = 0;
  if (my $oFile = PSOpen($path)) {
    if (PSSeek($oFile,0x210,0,PS_FILE_BEGIN)) {
      my $buffer = PSReadRaw($oFile,8);
      PSCloseHandle($oFile);
      my @vals = unpack("VV",$buffer);
      $t_val = getTime($vals[0],$vals[1]);
    }
    else {
      PSDisplayText("File seek to first offset failed.");
    }
  }
  else {
    PSDisplayText("File could not be opened.");
  }
  return gmtime($t_val);
```

```
}
#-----------------------------------------------------------------
# getRpDescr()
# Read the rp.log file to get the description and creation
# date
#-----------------------------------------------------------------
sub getRpDescr {
  my $path = shift;
  my $buffer;
  my $tag = 1;
  my $offset = 0x10;
  my @strs;
  my $str;
  my $oFile;
  if ($oFile = PSOpen($path)) {
    while ($tag) {
      PSSeek($oFile,$offset,0,PS_FILE_BEGIN);
      $buffer = PSReadRaw($oFile,2);
      if (unpack("v",$buffer) == 0) {
        $tag = 0;
      }
      else {
        push(@strs,$buffer);
      }
      $offset += 2;
    }
  }
  else {
    PSDisplayText("File could not be opened.");
  }
  PSClose($oFile);
  my $str = join('',@strs);
  $str =~ s/\00//g;
  return $str;
}
#-----------------------------------------------------------------
# getTime()
# Get Unix-style date/time from FILETIME object
# Input : 8 byte FILETIME object
# Output: Unix-style date/time
```

```
# Thanks goes to Andreas Schuster for the below code, which he
# included in his ptfinder.pl
#----------------------------------------------------------------
sub getTime {
  my $lo = shift;
  my $hi = shift;
  my $t;
  if ($lo == 0 && $hi == 0) {
    $t = 0;
  } else {
    $lo -= 0xd53e8000;
    $hi -= 0x019db1de;
    $t = int($hi*429.4967296 + $lo/1e7);
  };
  $t = 0 if ($t < 0);
  return $t;
}
```

When run, the sysrestore.pl ProScript starts by accessing the Registry within the image (this assumes that the examiner only has one system image open in ProDiscover, and has already populated the Registry View – see the ProDiscover instructions for how to populate the Registry View) in order to determine the path to the system root ("SystemRoot" is an environment variable within Windows that points to the Windows directory … on Windows XP, it is most often "C:\Windows"). From there, the ProScript populates the complete path to where the restore points are maintained (this value is stagnant or always in the same place on Windows XP systems), and begins parsing through the rp.log files within each restore point, and extracts information (timestamp, reason for the restore point being created) from the files.

Figure II.6 illustrates the output of the SysRestore.pl ProScript after it has been run against an image acquired from a Windows XP system.

**Figure II.6** Example output of the SysRestore.pl ProScript

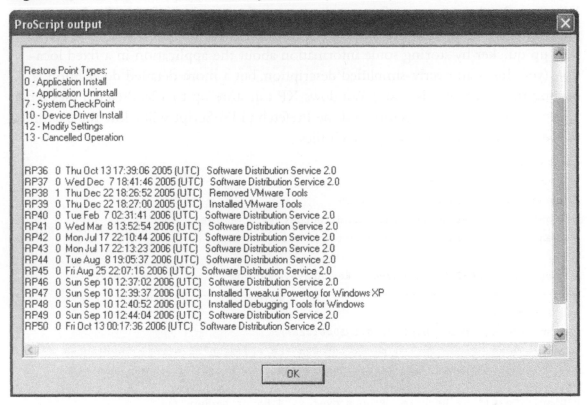

Figure II.6 clearly illustrates the utility of the SysRestore.pl ProScript. In the upper portion of the screen is a listing of the various codes that pertain to why a restore point is created. The rest of the output of the ProScript shows the restore points listed in sequential order, along with the code for the reason that the restore point was created, and the date (in UTC time) that the restore point was created. As you can see from Figure II.6, the system in question was most likely a Windows XP system running in a VMWare[11] session.

## Prefetch.pl

Besides System Restore Points, another interesting aspect of Windows XP systems is that, by default, Windows XP performs application prefetching.[12] Windows XP also

---

[11] http://www.vmware.com/

[12] http://technet.microsoft.com/en-us/library/bb457057.aspx

does boot prefetching, which Windows 2003 does by default, as well (although Windows 2003 does not perform application prefetching by default). The long and short of what application prefetching does is allow applications on Windows XP to start up quicker by storing some information about the application in a fixed location (yes, this is an overly-simplified description, but a more detailed description is beyond the scope of this book). Windows XP can store up to 128 Prefetch files (end in *.pf) in its Prefetch directory, and the Prefetch.pl ProScript will allow the examiner to see information about the prefetch files.

```perl
#! c:\perl\bin\perl.exe
#------------------------------------------------------------------
# Prefetch.pl, version 0.1_20061026
# ProScript to parse the Prefetch directory for .pf files, and
# then parse the files for run count and last run time.
#
# Copyright 2006-2007 H. Carvey, keydet89@yahoo.com
#------------------------------------------------------------------
use ProScript;
PSDisplayText("Prefetch.pl v. 0.1_20061026");
PSDisplayText("ProScript to parse through the Prefetch directory on Windows XP");
PSDisplayText("systems and return the filename, time last accessed, and the
run-count");
PSDisplayText(" -> Requires ProDiscover v. 4.85 or higher");
#------------------------------------------------------------------
# Get the SystemRoot value
my %sysinfo = ();
$numRegs = PSGetNumRegistries();

if ($numRegs == 0) {
    PSDisplayText("No registries to process");
    return;
}

$regName = PSGetRegistryAt(0);
PSRefreshRegistry($regName);
my $keyName = "HKEY_LOCAL_MACHINE\\Software\\Microsoft\\Windows NT\\
CurrentVersion";
my $rHandle = PSOpenRegistry($regName, $keyName);

if ($rHandle == 0) {
    PSDisplayText("Unable to locate registry key");
    return;
}
```

```perl
else {
#       PSDisplayText("Registry opened succesfully.");
}
# Access the key in order to get the SystemRoot value
while (1) {
    $RegKeyInfo = &ProScript::PSReadRegistry($rHandle);
  last if ($RegKeyInfo->{nType} == -1);
  next if ($RegKeyInfo->{nType} == PS_TYPE_KEY);
  my $value = $RegKeyInfo->{strRegName};
  my $data = $RegKeyInfo->{strValueData};
  $sysinfo{$value} = $data;
}
PSCloseHandle($rHandle);
#-------------------------------------------------------------
# Now we have a %sysinfo hash, and all we really want is the
# "SystemRoot" value
my $sysroot = $sysinfo{"SystemRoot"};
$sysroot = $sysroot."\\" unless ($sysroot =~ m/\\$/);
# Note: Make sure that the first letter (ie, the drive letter) of the
# SystemRoot path is capitalized; this is an issue with ProDiscover
$sysroot = ucfirst($sysroot);
my $objectName = PSGetObjectName(0);
my $path       = $objectName."\\".$sysroot."Prefetch";
my $pHandle = PSOpenDir($path,0);
if ($pHandle == NULL) {
    PSDisplayText("$path not opened.");
}

my $tag = 1;
while ($tag) {
    my $file = &ProScript::PSReadDirectory($pHandle);
    $tag = 0 if ($file == NULL || $file->{strName} eq "");
    my $pf = "pf";
    next if ($file->{bIsDirectory});
    next unless ($file->{strName} =~ m/$pf$/);
    my $filepath = $path."\\".$file->{strName};
    my ($t_val,$run);
    if (my $oFile = PSOpen($filepath)) {
      if (PSSeek($oFile,0x78,0,PS_FILE_BEGIN)) {
        my $buffer = PSReadRaw($oFile,8);
        my @vals = unpack("VV",$buffer);
        $t_val = getTime($vals[0],$vals[1]);
```

```perl
    }
    else {
       PSDisplayText("File seek to first offset failed.");
    }
# Get the Run count
    if (PSSeek($oFile,0x90,0,PS_FILE_BEGIN)) {
       my $buffer = &ProScript::PSReadRaw($oFile,4);
       PSCloseHandle($oFile);
       $run = unpack("V",$buffer);
    }
    else {
      PSDisplayText("File seek to second offset failed.");
    }
    PSDisplayText($file->{strName}." ".gmtime($t_val)." (UTC) ".$run);
  }
  else {
    PSDisplayText("File could not be opened.");
  }
}
PSCloseHandle($pHandle);
#----------------------------------------------------------------
# getTime()
# Get Unix-style date/time from FILETIME object
# Input : 8 byte FILETIME object
# Output: Unix-style date/time
# Thanks goes to Andreas Schuster for the below code, which he
# included in his ptfinder.pl
#----------------------------------------------------------------
sub getTime {
  my $lo = shift;
  my $hi = shift;
  my $t;

  if ($lo == 0 && $hi == 0) {
    $t = 0;
  } else {
    $lo -= 0xd53e8000;
    $hi -= 0x019db1de;
    $t = int($hi*429.4967296 + $lo/1e7);
  };
  $t = 0 if ($t < 0);
  return $t;
}
```

The Prefetch.pl ProScript is launched in much the same way as other ProScripts, and does not require any arguments. Once the ProScript locates the Prefetch directory, it begins parsing though each file, locating the timestamp for the last time the application was launched as well as the total number of times the application was launched, at specific, known offsets (0x78 and 0x90, respectively … these offsets are different for Prefetch files retrieved from Windows Vista sytems) within the binary file itself. As illustrated in Figure II.7, this information is then displayed in the ProScript output window.

**Figure II.7** Output of Prefetch.pl

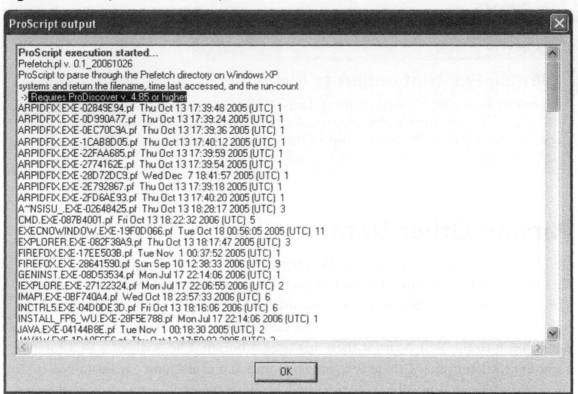

As you can see in Figure II.7, a great deal of useful information can be derived from the application prefetch files. However, the examiner should keep in mind that the Windows XP prefetch capability applies to all users on the system; in order to correlate running specific applications to a specific user, additional means of analysis (already discussed in this Part) will need to be employed.

> **NOTE**
>
> Windows Vista also maintains Prefetch files, as well. However, the offsets to the specific metadata maintained within the Prefetch file differ for Vista. The Prefetch.pl ProScript is meant only for Windows XP.

## Swiss Army Knife

### Writing ProScript output to files

Sometimes, trying to view a great deal of information in the ProScript display window can be quite cumbersome. As ProScripts are simply Perl scripts, code can be added to write the output of the ProScript to a file, as well as or instead of sending it straight to the display window.

# Parsing Other Data

Besides log files and binary data on Windows systems, there is quite a bit of other data that can be parsed in a number of useful ways. For example, the Visa Payment Card Industry (PCI) Data Security Standard (DSS)[13] has put forth requirements not only for notification of individuals in case their data has been compromised, but also notification to the PCI board if there has been a breach. The goals of a PCI forensic audit are to determine if there was, in fact, a breach and if credit card data was on the affected systems and possibly compromised. So what generally happens is that some systems may be forensically acquired, and the images will be analyzed for signs of an intrusion, as well as searched for credit card numbers. Forensic analysis tools such as EnCase provide the capability for the user to define a search for credit card numbers[14]

---

[13] http://www.corporate.visa.com/pd/security/main.jsp
[14] http://en.wikipedia.org/wiki/Credit_card_number

(or magnetic strip or "track" data) using a regular expression search, or for the analyst to use already-written scripts to perform the searches for them. With EnCase Forensic Edition version 5, for example, there is a Credit Card Finder module that is included with the Sweep Case EnScript (EnScripts are the user-definable scripting components of EnCase). Whether the analyst searches just for credit card numbers, or performs a more extensive search for track 1 or track 2 data (again, information which is maintained on the magnetic stripe of the credit card itself), she should not be surprised when the search returns hundreds or thousands (or tens of thousands) of hits.

The format of the credit card number is quite simple; generally 13 to 16 digits in length (some European cards can have 18 or 19 digits), and begin with certain sequences of numbers, referred to as the Bank Identification Number, or BIN. A description this general can (and will) return thousands of hits on almost any computer system, so we have to take another aspect of the credit card number into account. In order to be a valid credit card number, the number has to successfully pass a check via the Luhn Formula.[15] The Luhn Formula (or Algorithm) is a modulus 10 checksum that is used to validate credit card numbers. This is not a cryptographic hash function used for security purposes, but rather a checksum used for validation. Code to perform a Luhn algorithm verification looks like this:

```perl
sub luhn {
  my $num = shift;
  my $len = length($num);
  my @n;
  foreach (0..($len - 1)) {
    push(@n,substr($num,$_,1));
  }
  my @m = reverse @n;
  my @r;
  foreach my $i (0..($len - 1)) {
    if (($i % 2) == 0) {
      $r[$i] = $m[$i];
    }
    else {
      my $x = $m[$i] * 2;
      ($x > 9) ? ($r[$i] = $x - 9) : ($r[$i] = $x);
    }
```

---

[15] http://en.wikipedia.org/wiki/Luhn_algorithm

```
}
my $sum;
foreach my $i (0..($len - 1)) {
  $sum += $r[$i];
}
  my $v = $sum % 10;
  ($v == 0) ? (return 1) : (return 0);
}
```

The luhn() function takes a credit card number as it's argument, and processes it, verifying whether it is a valid credit card number or not. The function returns 1 if the number is a valid credit card number, 0 if it isn't. You will notice that the function makes no checks as to the length of the number, nor does it verify the BIN of the credit card number. EnCase's EnScript functionality includes a built-in string function called IsValidCreditCard() that perform these checks. I tend to use the luhn() code even when parsing though the search hits returned by EnCase when searching for credit card numbers, as an additional verification.

Perl code to validate or determine the type or BIN of the credit card number might look something like this:

```
sub getType {
  my $cc = shift;
  my $len = length($cc);
  return undef if ($len > 19);
  my $type;
  if ($len == 16) {
    if ($cc =~ m/^4/) {
      $type = "Visa";
    }
    elsif ($cc =~ m/^[51|52|53|54|55]/) {
      $type = "MasterCard";
    }
    elsif ($cc =~ m/^6011/) {
      $type = "Discover";
    }
    else {
      $type = "Unknown";
    }
  }
  elsif ($len == 15) {
    $type = "AmEx" if ($cc =~ m/^[34|37]/);
```

```
    }
    elsif ($len == 14) {
      $type = "Diners" if ($cc =~
m/^[36|38|300|301|302|303|304|305]/);
    }
    elsif ($len == 13) {
      $type = "Visa" if ($cc =~ m/^4/);
    }
    else {
      $type = "Unknown";
      print $cc." is ".$len." digits long.\n";
    }
}
```

As you can see, as with the luhn() function, the getType() function takes a single argument, which is the credit card number itself. The function checks the length and the BIN for the credit card number and returns either the type (i.e., "Visa", "MasterCard", etc.) or "Unknown". This makes things sorting the credit card numbers based on type (or BIN) a fairly straightforward process.

The track data is a bit different as it includes the credit card number (a.k.a., primary account number, or PAN). Track 1 data has the following format (keep in mind that this is a very simple overview):

```
B{PAN}~{Name}^{Other data}
```

Track 1 data starts with a format code of "B", and is followed by the PAN and the "^" separator. The Name field is 19 characters long and is made up of ASCII characters; this is followed by another "^" separator, and another 18 digits of data, the first four of which should be the expiration (expiry) date of the credit card. If you're parsing through a flat ASCII text file that contains one search hit for track 1 data on each line, you might parse through it using Perl code that looks like this (the scalar $file is the name of the file that holds our data):

```
my %cc_nums;
open(FH,"<",$file) || die "Could not open $file: $!\n";
while(<FH>) {
  chomp;
  next unless ($_ =~ m/^B/);
  my ($pan,$name,$rest) = split(/\x5e/,$_,3);
  $pan =~ s/[^0-9]//g;
  my $expiry = substr($rest,0,4);
```

```
   my $stuff = join(':',$pan,$name,$expiry);
   $cc_nums{$stuff} = 1;
   printf "%-20s %-20s %-4s\n",$pan,$name,$expiry;
}
close(FH);
```

The first thing this code snippet does is create a filehandle to our file (or "$file"), and opens the file in read-only mode. If for some reason the file can't be opened, the code will die() with an error message (hopefully) telling us why the file couldn't be opened. From there, we read the file a line at a time, chomp()'ing off the carriage return at the end of the line. If the line does not start with a "B" (remember, this is track 1 data), then we skip the line. From there, we split the line into components, based on the "∧" separator, denoted by its hexadecimal representation (\x5e). We then remove all non-numeric characters (between 0 and 9) from the PAN (there may be spaces or dashes), and then retrieve the expiration or expiry date from the remainder of the data in the third segment (i.e., the first four characters).

### Swiss Army Knife

### Valid Expiry Dates
The expiry date retrieved from the track 1 (and track 2) data consists of four digits in the form MMYY. I'll leave it as an exercise for you (the reader) to develop code to validate the expiry date.

Finally, the last thing the above code does is print out the parsed data in a fixed length format, using the printf() function. However, there's one other thing I'd like to point out, and that's the line that reads as follows:

```
$cc_nums{$stuff} = 1;
```

The purpose of this line is to remove duplicates. Many times when a search function goes through file (or acquired image), the search may return multiple instances of the same credit card number. This can happen for a number of reasons:

both track 1 and 2 data are found, the PAN exists in multiple files or multiple tables in a database, etc. Whatever the reason, you don't want to have to keep track of 10 copies of the same PAN. A real simple way to go about removing duplicates and guaranteeing uniqueness is to create a Perl hash, where the key to the hash is the data (PAN, or in the case of our code, PAN, name, and expiry date, separated by colons) we extracted, and we set the value for that key to 1 (it could be any value, really).

---

**NOTE**

David Schultze shared this technique for guaranteeing uniqueness in a dataset with me back in 1999 when we both worked at the same company. David shared a couple of really good programming tips with me, and this is one of them. I like to give credit where credit is due…thanks, David!

---

Track 2 data, on the other hand, has the following format:

```
{PAN}={Other data}
```

The track 2 data starts with the PAN (13–16 digits) and is followed by a "=" separator, and then up to 18 digits of additional data, the first four of which should (again) be the expiry date for the credit card. Perl code to parse through a flat ASCII text file containing a track 2 data search hit on each line might look similar to the following code:

```
my %cc_nums;
open(FH,"<",$file) || die "Could not open $file:
$!\n";
while(<FH>) {
  chomp;
  my ($pan,$rest) = split(/=/,$_,2);
  $pan =~ s/[^0-9]//g;
  my $expiry = substr($rest,0,4);
  my $stuff = join(':',$pan,$expiry);
  $cc_nums{$stuff} = 1;
}
close(FH);
```

This code has a lot of similarities to the code for parsing the track 1 data, so I won't go through it all again.

# Cc-sort.pl

Let's take a look at an example of using the Perl functions getType() and luhn().

```perl
use strict;
my $file = shift || die "You must enter a
filename.\n";
die "Could not find $file.\n" unless (-e $file);
open(FH,"<",$file) || die "Could not open $file: $!\n";
while(<FH>) {
  chomp;
  \isValid($_);
}
close(FH);

sub isValid {
  my $cc = shift;
# check cc number for validity; if the card is valid,
# the number and the type are returned
# strips out spaces and dashes
  $cc =~ s/[^0-9]//g;
# Verifies the length of the credit card number
  if (luhn($cc)) {
    my $type = getType($cc)) {
    print $type.":".$cc."\n";
  }
#       else {
#    print "Number is greater than 16 digits.\n";
#       }
}
```

This bit of code uses the luhn() and getType() functions to validate the credit card number, and then print out the type (BIN) of the credit card number, and the credit card number itself. This code serves as a very simple, yet straightforward example of how to use this code to validate credit card numbers located during a forensic investigation.

# Final Touches

Perl is an extremely useful and powerful tool for performing computer forensic analysis. While there are applications available that let an examiner access acquired images and perform some modicum of visualization, there are relatively few tools that meet the specific needs of a specific examiner working on a specific case. This is where the use of Perl really shines through and becomes apparent. For example, I received a request from another examiner not long ago, asking for some assistance in parsing a Windows Event Log file. I provided a copy of evt2xls.pl, and the examiner ran into issues with having far too many records in the resulting spreadsheet file for MS Excel to open. I made some quick changes to the script, and resent it…this led to a rather quick resolution of the issue, whereas prior to that, the examiner's ability to open the Event Log file and retrieve the necessary information was non-existent.

# Part III

# Monitoring Windows Applications with Perl

## Solutions for this Part:

- Core Application Processes
- Core Application Dependencies
- Network Connectivity
- Web Services
- Log Files

☑ Summary

# In This Toolbox

Working with enterprise-level Windows applications requires a great deal of analysis and constant monitoring. Automating the monitoring portion of this effort can save a great deal of time, reduce system downtimes, and improve the reliability of your overall application. By utilizing Perl scripts and integrating them with the application technology, you can easily build a simple monitoring framework that can alert you to current or future application issues.

In order to build this monitoring framework, you must first separate the individual components that make up your application, and determine a monitoring strategy for each. This allows you to build specific monitoring processes for each component while still providing a comprehensive view of the application as a whole from the monitoring perspective.

In this chapter, the components we will be focusing on are the core application processes, the core application dependencies, network connectivity, Web services, and log files. We will look at what the purpose of each component is and what its role is in the scope of the application as a whole. We will also break down each component even farther and look at what specific metrics we need to monitor or record to ensure that the component is functioning normally. Lastly, we will be putting together some scripts that monitor the application and build a complete monitoring framework for a sample application.

# Core Application Processes

The "core application processes" component refers to the basic executable(s) for the application that you are working with. Each executable takes up a certain amount of system resources including processor, memory, input/output (I/O), and so forth. By monitoring the state of the core system process and its use of system resources, you are able to determine how well the process is functioning and detect abnormalities.

In addition to the resources used by the application itself, we also need to be aware of resource use by other applications. Other applications that consume an extraordinary amount of system resources could have performance implications on the application we are monitoring, so it makes sense to stay aware of what these other applications are doing. We do this through monitoring of key performance indicators for the system and will be building some scripts for this purpose as well.

# Monitoring System Key Performance Indicators

Before we get to writing scripts to monitor our specific core processes, let us first take a look at the system as a whole. There are specific key performance indicators that can tell you the overall status of your system. Among these are processor utilization, memory utilization, and network utilization.

In this section, we will be creating scripts to watch each of these key performance indicators, measure their current status against a threshold, and take action when the threshold is exceeded. This will provide us some foundation scripts that we can then refine to provide information around our specific core processes.

## Monitoring System CPU Utilization

The system central processing unit (CPU) utilization is one of the most important metrics to monitor. When a system CPU is too busy, it is difficult for applications to get enough time slices with the CPU to perform their work. Consequently it is important to try to keep the CPU utilization down to a nominal level. This certainly doesn't mean that the process utilization should be low, however. That would mean that you spent too much money on hardware that you do not need. A good balance means keeping the system busy, but ensuring that there is enough processor capacity to handle normal loads and some peaks. We will look at exactly what these numbers should be when we look at thresholds.

First, let's put together a quick Perl script for getting the system processor utilization data. For all of these examples, I will be using Microsoft Windows 2003 Server. I will also be using Perl 5.8.8 as included in ActivePerl 5.8.8.882. As we go through the exercise of creating each of these scripts, a variety of Perl modules may be needed. As we get to each module, I will include the module name and a note indicating that you will need to install the referenced module.

In the case of working with system CPU utilization, there is a module available for Perl called Win32::PerfMon that allows for access to all data available through Windows Performance Monitor. This module should work very well for our purposes as we will eventually need a lot of data that can be provided through Performance Monitor. Unfortunately, at this time the module we need cannot be easily installed using Perl Package Manager (PPM) and does in fact require some additional work to compile, install, and configure. The module itself can be downloaded from CPAN at http://search.cpan.org/~glensmall/Win32-PerfMon-0.07/PerfMon.pm and is currently at version 0.07.

In order to compile this module, you must first have Microsoft Visual C++ 6.0 or later. In these examples, I will be using Microsoft Visual C++ 2005 Express Edition. This is a free download from Microsoft and is available at http://msdn2.microsoft. com/en-us/express/aa975050.aspx. Download and install the Visual C++ Express Edition prior to installing ActivePerl if possible.

In addition, you will also need the Microsoft Platform SDK. This is a development kit that allows you to create applications using functions in Microsoft's platform libraries. The SDK is available at http://www.microsoft.com/downloads/details. aspx?familyid=0baf2b35-c656-4969-ace8-e4c0c0716adb&displaylang=en. Download the SDK and install it using the standard installer.

```
Post-installation, there are several steps you must take to get the SDK to work
properly with Visual C++ Express Edition. First, open up the C:\Program Files\
Microsoft Visual Studio 8\VC\vcpackages\VCProjectEngine.dll.express.config file.
Modify the "Directories" section as shown below:
  Include="$(VCInstallDir)include;$(VCInstallDir)PlatformSDK\include;$(FrameworkSDKDir
)include;C:\Program Files\Microsoft Platform SDK for Windows Server 2003 R2\Include"

  Library="$(VCInstallDir)lib;$(VCInstallDir)PlatformSDK\lib;$(FrameworkSDKDir)lib;
$(VSInstallDir);$(VSInstallDir)lib;C:\Program Files\Microsoft Platform SDK for
Windows Server 2003 R2\Lib"

  Path="$(VCInstallDir)bin;$(VCInstallDir)PlatformSDK\bin;$(VSInstallDir)Common7\
Tools\bin;$(VSInstallDir)Common7\tools;$(VSInstallDir)Common7\ide;$(ProgramFiles)\
HTML Help Workshop;$(FrameworkSDKDir)bin;$(FrameworkDir)$(FrameworkVersion);
$(VSInstallDir);C:\Program Files\Microsoft Platform SDK for Windows Server 2003
R2\Bin;$(PATH)"
```

Basically, you are adding the Platform SDK paths to the "Include," "Library," and "Path" variable settings. Next, you'll need to delete the \%USERPROFILE%\Local Settings\Application Data\Microsoft\VCExpress\8.0\vccomponents.dat file as it caches these settings. The next step enables the compiler dependencies to work correctly. Modify the *C:\Program Files\Microsoft Visual Studio 8\VC\VCProjectDefaults\ corewin_express.vsprops* file and change the AdditionalDependencies value to:

```
AdditionalDependencies="kernel32.lib user32.lib gdi32.lib winspool.lib comdlg32.lib
advapi32.lib shell32.lib ole32.lib oleaut32.lib uuid.lib"
```

We're almost done with these necessary changes. The last requirement is that you modify the paths used by Visual C++ in the *C:\Program Files\Microsoft Visual Studio 8\Common7\Tools\vsvars32.bat* file. Change the "PATH," "INCLUDE," and "LIB" lines as shown below:

```
@set PATH=C:\Program Files\Microsoft Visual Studio
8\Common7\IDE;C:\Program Files\Microsoft Visual Studio
```

```
8\VC\BIN;C:\Program Files\Microsoft Visual Studio
8\Common7\Tools;C:\Program Files\Microsoft Visual Studio
8\SDK\v2.0\bin;C:\WINDOWS\Microsoft.NET\Framework\v2.0.50727;
C:\Program Files\Microsoft Visual Studio
8\VC\VCPackages;C:\Program Files\Microsoft Platform SDK for
Windows Server 2003 R2\Bin;%PATH%

@set INCLUDE=C:\Program Files\Microsoft Visual Studio
8\VC\INCLUDE;C:\Program Files\Microsoft Platform SDK for
Windows Server 2003 R2\Include;%INCLUDE%

@set LIB=C:\Program Files\Microsoft Visual Studio
8\VC\LIB;C:\Program Files\Microsoft Visual Studio
8\SDK\v2.0\lib;C:\Program Files\Microsoft Platform SDK for
Windows Server 2003 R2\Lib;%LIB%
```

Those are all the required changes, but there is one additional change necessary if you would like to use the Visual C++ GUI to create Windows applications. This is not required for compiling the Win32::PerfMon module, but may be useful to you. Edit the *C:\Program Files\Microsoft Visual Studio 8\VC\VCWizards\AppWiz\Generic\Application\ html\1033\AppSettings.htm* file. On lines 441 through 444, put a "//" at the beginning of the lines to comment them out. This should look like the following when complete:

```
// WIN_APP.disabled = true;
// WIN_APP_LABEL.disabled = true;
// DLL_APP.disabled = true;
// DLL_APP_LABEL.disabled = true;
```

After installing these packages and prior to working with Perl, open a command window and change to the binaries directory used for Visual C++ Express Edition (such as *C:\Program Files\Microsoft Visual Studio 8\VC\bin\*) and run the batch file vcvars. bat. This batch file sets the environment variables used by Visual C++ Express Edition.

After loading the environment variables, do not close the command prompt. These will be necessary for use with Perl when compiling the Win32::PerfMon module. Assuming that you have Perl in your path, change to the directory where you downloaded the Win32::PerfMon module. Decompress the module and change into the Win32-PerfMon-0.07 directory. Run the following commands to compile and install the module:

```
perl Makefile.PL
nmake -f Makefile all
nmake install
```

The first command uses Perl to create all the necessary Makefile data for compiling the module. Next we run the "nmake" utility, which is part of Visual C++. This utility uses the Makefile to compile the module. Finally, we install the module into the correct location for use by Perl through another use of the nmake utility. After completing these steps, you should have a screen that looks similar to that shown in Figure III.1.

**Figure III.1** Win32::PerfMon Installation

Lastly, due to some changes in the way that the 2005 edition of Visual C++ works, we have to change the manifest data for the *PerfMon.dll* file. This is unnecessary if you are using Visual C++ 6.0. Change to the *C:\Perl\site\lib\auto\Win32\PerfMon* directory in your command window and run the following command:

```
mt /manifest PerfMon.dll.manifest /outputresource:PerfMon.dll;#2
```

This basically embeds the generated manifest file into the dynamic link library (DLL) as a resource allowing the DLL used by the Win32::PerfMon module to work properly. Again, this is only necessary if you are using the Visual C++ Express Edition 2005 compiler to compile the module.

With this module installed, we now have access to all of the Performance Monitor data through the use of the following line in our Perl script:

```
use Win32::PerfMon;
```

So for our CPU utilization script, this will obviously play a critical role. What we want to do first is gather the CPU utilization data. After we are able to collect the information, we'll figure out how to use it as a data point for actual monitoring activities.

To collect CPU utilization, we will be using the performance monitor counter "% Processor Time." According to the description of this counter from Microsoft, "% Processor Time is the percentage of elapsed time that the processor spends to execute a non-idle thread. It is calculated by measuring the duration of the idle thread is active in the sample interval, and subtracting that time from interval duration. (Each processor has an idle thread that consumes cycles when no other threads are ready to run). This counter is the primary indicator of processor activity, and displays the average percentage of busy time observed during the sample interval. It is calculated by monitoring the time that the service is inactive, and subtracting that value from 100 percent.

Let's put together some code for this. What we'll do is use the Win32::PerfMon module and capture the data available in the "% Processor Time" counter. This is part of the "Processor" performance monitor object. This object has multiple instances available depending on the number of processors in the system. If only one processor exists, the data is stored under instance 0. There is also a special instance available for objects such as this, which provides the data from all other instances. This is the "_Total" instance. In most cases, you will not need to know the performance data for a specific processor when monitoring an application, so we'll use the "_Total" instance for our data. The code to pull this data is shown in Figure III.2.

## Figure III.2 get_percentprocessortime.pl

```perl
#get_percentprocessortime.pl
use Win32::PerfMon;
use strict;

my $ret = undef;
my $err = undef;
my $Object = undef;
my $Counter = undef;
my $CounterData = undef;
#connect to localhost for data
my $perfmon = Win32::PerfMon->new("\\\\localhost");

if($perfmon != undef)
{
    $ret = $perfmon->AddCounter("Processor",
                    "% Processor Time", "_Total");
    if($ret != 0)
    {
        $ret = $perfmon->CollectData();
        if($ret  != 0)                              {
            my $proctime = $perfmon->GetCounterValue(
                "Processor","% Processor Time",
                "_Total");
            if($proctime > -1)
            {
            print "% Processor Time = [$proctime]\n";
            }
            else
            {
            $err = $perfmon->GetErrorText();
            print "Failed to get the counter data!\n",
                $err, "\n";
            }
        }
        else
        {
            $err = $perfmon->GetErrorText();
            print "Failed to collect the perf data!\n",
                $err, "\n";
        }
    }
    else
    {
        $err = $perfmon->GetErrorText();
        print "Failed to add the counter!\n", $err, "\n";
    }
}
else
{
    print "Failed to create the perf object!\n";
}
```

The example shown above (and all future examples in this chapter) may not be formatted in strict adherence to the rules of "readable code," due to the column limitations of the print format. In this code, we're doing a lot of error detecting. A lot can go wrong when pulling the performance data ranging from being unable to connect to the host system to being unable to get data from a specific counter. To make strong code, we need to capture as many of these errors as possible and do something with them. The Win32::PerfMon module makes this easy by returning a "−1" whenever it is unable to perform a specific function. By noting the place where we receive this return code, we can determine to some degree what went wrong.

Walking through the code, the first active thing that we do is create an object using the Win32::PerfMon module. We'll name that object *$perfmon*. Next, we'll add a counter to the object by calling the *AddCounter* function and pointing it to the counter we're interested in. The syntax of this is to specify the performance monitor object, then the counter, then the instance. Next, we collect the data from the performance monitor using the *CollectData* function. This takes a little over one second to gather. While technically the poll can be accomplished in less time than this, some counters require that a one-second delay be instantiated between two data requests in order to properly collect the values. The PerfMon module takes this into account and collects the data, sleeps for one second, then collects the data a second time.

After the data is collected, we then must get the value by using the *GetCounterValue* function. This data is then stored in the *$proctime* variable and displayed using the *print* function. If we encounter any errors throughout this process, we display an error message followed by the error that the module returns by using the *GetErrorText* function.

## Monitoring System Memory Utilization

The next statistic that we'll monitor is memory utilization for the system. Again, we'll use the Win32::PerfMon module to gather this data. Since the system CPU utilization and the system memory utilization are both statistics of the same type (both are whole–system related), we can gather them both in the same script and still adhere to a modular architecture. Figure III.3 shows an example of how this can be accomplished.

## Figure III.3 get_system_stats.pl

```perl
#get_system_stats.pl
use Win32::PerfMon;
use strict;

my $ret = undef;
my $err = undef;
my $Object = undef;
my $Counter = undef;
my $CounterData = undef;
#connect to localhost for data
my $perfmon = Win32::PerfMon->new("\\\\localhost");

if($perfmon != undef)
{
    $ret = $perfmon->AddCounter("Processor",
            "% Processor Time", "_Total");
    if($ret != 0) {
        $ret = $perfmon->AddCounter("Memory",
                "Available MBytes", -1);
    }
    if($ret != 0)
    {
        $ret = $perfmon->CollectData();
        if($ret  != 0)                  {
            my $proctime = $ perfmon->GetCounterValue(
                    "Processor","% Processor Time",
                    "_Total");
            if($proctime > -1)
            {
            print "% Processor Time = [$proctime]\n";
            }
            else
            {
            $err = $perfmon->GetErrorText();
            print "Failed to get the counter data!\n",
                    $err, "\n";
            }
            my $freemem = $perfmon->GetCounterValue(
                    "Memory","Available MBytes",
                    -1);
            if($freemem > -1)
            {
            print "Available Memory = [$freemem]MB\n";
            }
            else
            {
            $err = $perfmon->GetErrorText();
            print "Failed to get the counter data!\n",
                    $err, "\n";
            }
        }
        else
        {
            $err = $perfmon->GetErrorText();
            print "Failed to collect the perf data!\n",
                    $err, "\n";
```

```
            }
        }
        else
        {
            $err = $perfmon->GetErrorText();
            print "Failed to add the counter!\n", $err, "\n";
        }
    }
    else
    {
        print "Failed to create the perf object!\n";
    }
}
```

In this case, we check to make sure that the previous counter addition was successful and then add another counter. The counter we're adding uses the "Memory" object and the "Available MBytes" counter. There are no individual instances available for this object and counter, so we use the −1 value as the instance to indicate this.

If all counters were successfully added, we move on to collecting the data. Again, assuming that the collection went well (and capturing errors as needed when it isn't), we then display the data. Now we have two critical system statistics: % Processor Time and Available MBytes.

## Monitoring System Network Utilization

The last system-wide statistic that we'll look at is network utilization. There are many other system statistics that you as an application administrator may be interested in, but the selection we have included here focuses on those counters that are common requirements for monitoring every application. I feel it's important to note that we will be covering dependencies such as disk utilization later in the chapter rather than here within the KPI section.

Monitoring network utilization on a Windows server takes a little bit more work than the other statistics that we have gathered so far. Some additional calculations are required in order to obtain useful information out of the data provided by the performance monitor counter. The figures we need for these calculations are both provided by the "Network Interface" object. We'll be using the "Bytes Total/sec" and "Current Bandwidth" counters. The instance name will be the name of the network adapter that you are working with.

To determine the actual amount of network utilization, we need to multiply the "Bytes Total/sec" value by 8, and then divide the result by the value of the "Current Bandwidth" counter. This will provide us with an overall usage percentage. The code shown in Figure III.4 shows an example of how this can be done.

## Figure III.4 get_system_stats.pl

```perl
#get_system_stats.pl
use Win32::PerfMon;
use strict;

my $ret = undef;
my $err = undef;
my $Object = undef;
my $Counter = undef;
my $CounterData = undef;
#connect to localhost for data
my $perfmon = Win32::PerfMon->new("\\\\localhost");

if($perfmon != undef)
{
    $ret = $perfmon->AddCounter("Processor",
                "% Processor Time", "_Total");
    if($ret != 0) {
        $ret = $perfmon->AddCounter("Memory",
                "Available MBytes", -1);
    }
    if($ret != 0) {
        $ret = $perfmon->AddCounter("Network Interface",
                "Bytes Total/sec",
                "Intel 21140-Based PCI Fast ".
                "Ethernet Adapter [Generic]");
    }
    if($ret != 0) {
        $ret = $perfmon->AddCounter("Network Interface",
                "Current Bandwidth",
                "Intel 21140-Based PCI Fast ".
                "Ethernet Adapter [Generic]");
    }
    if($ret != 0)
    {
        $ret = $perfmon->CollectData();
        if($ret != 0)            {
            my $proctime = $perfmon->GetCounterValue(
                "Processor","% Processor Time",
                "_Total");
              if($proctime > -1)
            {
            print "% Processor Time = [$proctime]\n";
            }
            else
            {
            $err = $perfmon->GetErrorText();
            print "Failed to get the counter data!\n",
                $err, "\n";
            }

            my $freemem = $perfmon->GetCounterValue(
                "Memory","Available MBytes",
                -1);
            if($freemem > -1)
            {
            print "Available Memory = [$freemem]MB\n";
            }
```

```perl
    else
    {
    $err = $perfmon->GetErrorText();
    print "Failed to get the counter data!\n",
          $err, "\n";
    }

    my $netbps = $perfmon->GetCounterValue(
          "Network Interface",
          "Bytes Total/sec",
          "Intel 21140-Based PCI Fast ".
          "Ethernet Adapter [Generic]");
    if($netbps > -1)
    {
    print "Network Bytes/Second = [$netbps]\n";
    }
    else
    {
    $err = $perfmon->GetErrorText();
    print "Failed to get the counter data!\n",
          $err, "\n";
    }

    my $netbw = $perfmon->GetCounterValue(
          "Network Interface",
          "Current Bandwidth",
          "Intel 21140-Based PCI Fast ".
          "Ethernet Adapter [Generic]");
    if($netbw > -1)
    {
    print "Network Bandwidth Bits/Second =".
          " [$netbw]\n";
    }
    else
    {
    $err = $perfmon->GetErrorText();
    print "Failed to get the counter data!\n",
          $err, "\n";
    }

    if ($netbps > -1 && $netbw > -1)
    {
            my $netutil = sprintf "%.2f",
                  (8 * $netbps / $netbw);
            print "% Network Utilization = ".
                  " [$netutil]\n";
    }
    else
    {
            print "Not enough data to determine".
                  "network utilization!";
    }
}
```

```
            else
            {
                    $err = $perfmon->GetErrorText();
                    print "Failed to collect the perf data!\n",
                            $err, "\n";
            }
        }
        else
        {
            $err = $perfmon->GetErrorText();
            print "Failed to add the counter!\n", $err, "\n";
        }
    }
    else
    {
        print "Failed to create the perf object!\n";
    }
```

In this example, you can see that we're continuing to add the new counters in the same manner as we used for the previous counter addition. Then we're displaying the new network statistics as usual. What changes is that we're doing a quick check to ensure that there is data in those variables and then performing the necessary calculation and returning the result. This allows us to provide the network utilization statistic if we can, and handle the error message properly if we can't.

In addition, we are doing some formatting of the network utilization data. The following line of code demonstrates one method of doing this formatting.

```
my $netutil = sprintf "%.2f", (8 * $netbps / $netbw);
```

By using the "sprintf" function, we are able to round the percentage of network utilization to two decimal places so that the number is useable. It's very difficult to work with numbers such as "5.84e–006" from a readability perspective, and this is the type of number that could be returned if we didn't do some rounding and formatting to get the result.

## Swiss Army Knife

### Using Performance Monitor Data

In writing these scripts, we have barely touched the surface of the data provided through performance monitor. Using the Windows Resource Kit utility "typeperf," you can get information on which counters are available from a command-line interface. This is much faster than browsing through the PerfMon utility itself.

Running the command "typeperf–q" will list all counters. You can further refine your counter search by specifying objects to list the counters for. Using the command "typeperf-?" will print out help for the tool.

# Monitoring a Core Application Process

Now that we have some information on the system, we can move on to monitoring some portions of the application. We'll start with the core application process and its availability as well as its use of system resources. As an application can have multiple processes running that provide different functions, we'll refer to the process that serves the purpose of being the parent to the other processes as the *core process*, with any other processes spawned from the parent considered *dependent processes*.

## Monitoring Process Availability a Specific Process

One of the most important things to monitor for a core process is to make sure that the process is actually up and running. Statistics such as CPU and memory utilization are very important, but they won't help if the process isn't running. So the first thing we want to monitor is the availability of the core process itself.

In this case, our best bet for obtaining good process information is to use the Win32::Process::Info module. This module can be loaded through PPM and can be found in the default ActiveState repository. Figure III.5 shows this module through the graphical user interface (GUI) version of PPM.

**Figure III.5** Win32::Process::Info Module

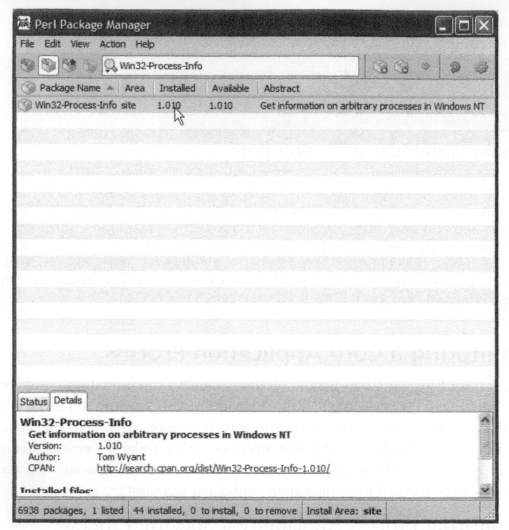

The script in Figure III.6 shows how this module can be used to gather information on the processes running on the system. Keep in mind that at this point we are not looking for performance information; rather we are looking for details on the processes themselves and their availability. First, we'll get a list of processes running on the system.

**Figure III.6** get_processes.pl

```perl
#get_processes.pl
use strict;
use Win32::Process::Info;
my $pi = Win32::Process::Info->new();
my @procinfo = $pi->GetProcInfo();
for my $pid (@procinfo){
    print $pid->{"ProcessId"}." ".
        $pid->{"Name"}."\n";
}
```

In the code sample in Figure III.6, we first create a new instance of the Win32::
Process:: Info object and call it $pi. Next, we call the "GetProcInfo" function of the
module to gather the process information into the "procinfo" list. Cycling through
each member of the list, we then print the process ID and name for each running
process by displaying the "ProcessId" and "Name" keys associated with the process.

Now that we have a way to query for the processes running on a system, we
need to work with that data and find a single specific process that we are interested
in. The "GetProcInfo" function, when called with no parameters, loads all of the data
that it obtains into a list of anonymous hashes. Using this list, we can find a process
by either process ID or executable name by comparing the value we are looking for
against the "ProcessId" key or "Name" key, respectively.

Some programs, when started, create a "PID" file containing the process ID for
the process being run. While this is more common on *NIX operating systems, some
Windows applications use the same standard. If this is the case, the application typically
creates a file called *<application name>.pid* in the working directory for the application.
Reading in the value stored in this file allows us to obtain the specific process ID and
use it to get the process status. The code shown in Figure III.7 shows how this file
could be read and the value utilized to check on the process.

**Figure III.7** get_process_status_id.pl

```
#get_process_status_id.pl
use strict;
use Win32::Process::Info;
my $pi = Win32::Process::Info->new();
my @procinfo = $pi->GetProcInfo();
my $procstatus=0;

my $result = open PIDFILE, "C:\\dls\\Komodo.pid";
if ($result) {
    if ( defined(my $piddata = <PIDFILE>)) {
        chomp $piddata;
        for my $pid (@procinfo){
            if ($pid->{"ProcessId"} == $piddata) {
                print $pid->{"Name"}." is running!";
                $procstatus=1;
            }
        }
        if ($procstatus!=1)
        {
            print "Process $piddata cannot be found!"
        }
    }
    else
    {
        print "PID not found in PID file.";
    }
}
else
{
    print "PID file not found.";
}
```

In the code shown in Figure III.7, we use the Win32::Process::Info module in a manner similar to that in Figure III.6, but add in some additional features. The first major change is the definition of the PID file, opening the file, and reading in the process ID from the first line of the file. We also use the "chomp" function to get rid of any new line characters. Then we cycle through all of the hashes provided by the "GetProcInfo" function looking for the process ID gathered from the PID file. If this is found, we note that the process is running (including the executable name) and change the $procstatus variable to contain a value of 1. Later, we check this variable and if the value is not 1, we display a message stating that the process is not found. This implies that the process is either not running or that the PID information is incorrect. Consequently, the process ID gathered from the PID file is included in the message for validation purposes.

Using this code, we can also find a process by using its name. This allows us additional flexibility in our monitoring so that we are not reliant on applications creating their own PID files. Figure III.8 shows an example of how this code can be modified to check the availability of a process using its name rather than its process ID.

### Figure III.8 get_process_status_name.pl

```perl
#get_process_status_name.pl
use strict;
use Win32::Process::Info;
my $pi = Win32::Process::Info->new();
my @procinfo = $pi->GetProcInfo();
my $procstatus=0;
my $procname="komodo.exe";

for my $pid (@procinfo){
    if ($pid->{"Name"} eq $procname) {
        print $pid->{"Name"}." is running under ".
            "process id ".$pid->{"ProcessId"}."!";
        $procstatus=1;
    }
}
if ($procstatus!=1)
{
    print "Process $procname cannot be found!"
}
```

This is very similar code to that shown in Figure III.7. The differences are the removal of the file operations used for the PID file and changing the search to use the "Name" key rather than the "ProcessId" key. This allows us to perform a string comparison operation using the defined process name rather than its process ID. The $procstatus variable is used in the same manner as it was in the Figure III.7 example.

## Swiss Army Knife

### Monitoring Multiple Processes

You can use the script shown in Figure III.7 or the other process-specific scripts to monitor multiple processes. Just gather information from multiple PID files or use multiple process names when gathering your statistical data. For best performance of the monitoring script, gather all of the process-specific data first, then go through it and display or work with the data as needed.

# Monitoring CPU Utilization for a Specific Process

Collecting the overall processor utilization for the system is very important, but in many cases we may be concerned about the utilization by a single process for the application that we are monitoring. So let's modify the script we have developed for monitoring CPU utilization shown in Figure III.2 by adding the counter for a specific process.

In order to gather the CPU statistics for the process, we will need to know what the process name is. One method of finding the process name based on a process ID is shown in Figure III.7. In this example, we'll assume that we are working with a PID file as shown in Figure III.7.

The counter to use for finding the CPU utilization for a specific process can be found by using the "Process" object instead of the "Processor" object that we used previously. We can then use the "% Processor Time" counter again, since that's the statistic we are interested in. And then, when specifying the instance, just use the name of the executable that you want to get information on. In this case, we'll be getting information on the Perl editor Komodo Edit, which was also used for the prior examples. This can be seen in Figure III.9.

## Figure III.9 get_percentprocessortime_komodo

```perl
#get_percentprocessortime_komodo.pl
use Win32::PerfMon;
use strict;
use Win32::Process::Info;
my $pi = Win32::Process::Info->new();
my @procinfo = $pi->GetProcInfo();
my $procname = undef;
my $procstatus=0;
my $ret = undef;
my $err = undef;
my $Object = undef;
my $Counter = undef;
my $CounterData = undef;
#connect to localhost for data
my $perfmon = Win32::PerfMon->new("\\\\localhost");

my $result = open PIDFILE, "C:\\dls\\Komodo.pid";
if ($result) {
     if ( defined(my $piddata = <PIDFILE>)) {
     chomp $piddata;
     for my $pid (@procinfo){
          if ($pid->{"ProcessId"} == $piddata) {
               $procname = $pid->{"Name"};
               print $pid->{"Name"}." is running!\n";
               $procstatus=1;
          }
     }

     if ($procstatus!=1)
     {
          print "Process $piddata cannot be found!"
     } else {
          $procname =~ s/^(.+?)(\.[^.]*)?$/$1/;
          if($perfmon != undef) {
               $ret = $perfmon->AddCounter("Process",
                    "% Processor Time", $procname);
               if($ret != 0) {
                    $ret = $perfmon->CollectData();
                    if($ret   != 0) {
                    my $proctime=$perfmon->GetCounterValue(
                         "Process","% Processor Time",
                         $procname);
                         if($proctime > -1) {
                         print "% Processor Time = ".
                              "[$proctime]\n";
                         } else {
                         $err = $perfmon->GetErrorText();
                         print"Failed to get the counter".
                              " data!\n", $err, "\n";
                         }
                    } else {
                         $err = $perfmon->GetErrorText();
                         print "Failed to collect the ".
                              "perf data!\n", $err, "\n";
```

```
        }
                        } else {
                                $err = $perfmon->GetErrorText();
                                print "Failed to add the counter!\n",
                                        $err, "\n";
                        }
                } else {
                        print "Failed to create the perf object!\n";
                }
        } else {
                print "PID not found in PID file.";
        }
} else {
        print "PID file not found.";
}
```

As you can see, we're doing practically the same thing as we did in our previous code sample in Figure III.2, but referring to the new object and a new instance. In addition, much of the code used for opening a PID file and finding the process name demonstrated in Figure III.7 is incorporated in this example as well. One additional line regarding the process name should be noted. The following statement is used to format the resulting process name and remove the "." and file extension:

```
$procname =~ s/^(.+?)(\.[^.]*)?$/$1/;
```

We didn't necessarily have to create a new PL file for this. We could have used the existing file "get_percentprocessortime.pl" and added a second counter. The reason that I broke this out separately is because for the overall architecture for our monitoring application, we should modularize certain sections for ease of use and maintenance.

In some cases, it makes sense to combine certain statistics into a single script as you saw in the example shown in Figure III.3. The general rule is to combine code when the objects that they act upon are similar (e.g., grouping all system-related code together or all core process code together). In this instance, however, one of these statistics is related to the system as a whole and the other to an individual process, therefore it makes sense architecturally to separate them. On the other hand, it does make sense to combine this code with the availability check for the individual process, so this example serves both purposes.

# Monitoring Memory Utilization for a Specific Process

Next, we'll be looking at monitoring the memory utilization for a specific process. Again, it makes sense to combine this effort with the other information we're gathering from the process CPU utilization. So we'll add a counter for the "Process" object, "Private Bytes" counter, and an instance based on the process ID shown in our PID file. Keep in mind that for any of these examples that use a PID file, you can easily convert them to use the process name in a manner similar to that shown in Figure III.8.

This particular counter is described as, "Private Bytes is the current size, in bytes, of memory that this process has allocated that cannot be shared with other processes" by Microsoft. This should fit our needs and provide useful information for our monitoring. The code for this is shown in Figure III.10.

## Figure III.10 get_process_stats.pl

```perl
#get_process_stats.pl
use Win32::PerfMon;
use strict;
use Win32::Process::Info;
my $pi = Win32::Process::Info->new();
my @procinfo = $pi->GetProcInfo();
my $procname = undef;
my $procstatus=0;
my $ret = undef;
my $err = undef;
my $Object = undef;
my $Counter = undef;
my $CounterData = undef;
#connect to localhost for data
my $perfmon = Win32::PerfMon->new("\\\\localhost");

my $result = open PIDFILE, "C:\\dls\\Komodo.pid";
if ($result) {
    if ( defined(my $piddata = <PIDFILE>)) {
    chomp $piddata;
    for my $pid (@procinfo){
        if ($pid->{"ProcessId"} == $piddata) {
            $procname = $pid->{"Name"};
            print $pid->{"Name"}." is running!\n";
            $procstatus=1;
        }
    }

    if ($procstatus!=1)
    {
        print "Process $piddata cannot be found!"
    } else {
        $procname =~ s/^(.+?)(\.[^.]*)?$/$1/;
```

```perl
            if($perfmon != undef) {
                $ret = $perfmon->AddCounter("Process",
                    "% Processor Time", $procname);
                if($ret != 0) {
                    $ret = $perfmon->AddCounter("Process",
                        "Private Bytes", $procname);
                }
                if($ret != 0) {
                    $ret = $perfmon->CollectData();
                    if($ret  != 0) {
                    my $proctime=$perfmon->GetCounterValue(
                        "Process","% Processor Time",
                        $procname);
                        if($proctime > -1) {
                        print "% Processor Time = ".
                            "[$proctime]\n";
                        } else {
                        $err = $perfmon->GetErrorText();
                        print"Failed to get the ".
                        "processor counter data!\n",
                        $err, "\n";
                        }
                    my $freemem = $perfmon->GetCounterValue(
                        "Process","Private Bytes",
                        "komodo");
                        if($freemem > -1)
                        {
        $freemem =~ s/(?<=\d)(?=(?:\d\d\d)+\b)/,/g;
        print "Memory used by process = [$freemem]" .
            " Bytes\n";
                        } else {
                        $err = $perfmon->GetErrorText();
        print "Failed to get the memory ".
            "counter data!\n",
            $err, "\n";
                        }
                    } else {
                        $err = $perfmon->GetErrorText();
                        print "Failed to collect the ".
                            "perf data!\n", $err, "\n";
                    }
                } else {
                    $err = $perfmon->GetErrorText();
                    print "Failed to add the counter!\n",
                        $err, "\n";
                }
            } else {
                print "Failed to create the perf object!\n";
                }
            }
    } else {
        print "PID not found in PID file.";
    }
} else {
    print "PID file not found.";
}
```

In this example, we modified the get_percentprocessortime_komodo.pl in a manner similar to what we did to combine the system statistics. Again, we're handling error messages in an appropriate manner and displaying the data appropriate to our specific process. One additional line of code in here that bears some attention is this one:

```
$freemem =~ s/(?<=\d)(?=(?:\d\d\d)+\b)/,/g;
```

In this case, we're formatting the data shown in the $freemem variable and storing it back in that variable. Why? Simply because it looks better when you're displaying a large figure if you add commas to separate values. Since this value does not have a decimal point, we can quickly format it and redisplay the value in a "prettier" manner.

It should also be noted that the data gathered from this performance counter is different from that shown in Windows Task Manager. This is not a bug, but a difference in the way that memory use is calculated. Task Manager adds together the memory allocated to the specific process exclusively with the memory shared with other processes, and displays that cumulative figure. This performance monitor counter only shows the memory dedicated specifically to the process. In truth, neither method is 100 percent accurate, but both give an estimate that is close enough to use for monitoring purposes. Just be sure that you understand that statistics pulled in one manner may not necessarily match those pulled in another.

# Setting and Using Thresholds

When monitoring an application, ideally you want to be able to take action based on the data that you receive. For example, you may want to restart a process if it goes down. Or perhaps send out an e-mail when your system reaches a high percentage of utilization. In monitoring terms, thresholds are used for determining when these actions should be taken.

So far we've written several scripts that allow us to gather important data about core processes and the system itself, but we're not doing anything with that data other than displaying it. In order to take action based on the data, we need do the following:

- Define which values are okay for each piece of data
- Define which values are bad for each piece of data
- Define which values are critically bad for each piece of data

These can be defined as status colors such as green, yellow, and red, or status codes such as good, warning, and critical. Since we'll be defining thresholds for a lot of different pieces of data that we're obtaining, we need a generic manner of handling the evaluation of the threshold and taking action on the threshold. The best method of doing this is to create a separate script just for dealing with threshold-related items.

In this script, we need to be able to take incoming information on what is being monitored and what the resulting value of the monitor is. Then we'll need to compare that value with a set of threshold values and see how they relate. Finally, based on the result, we'll need to take action of some type even if that action is to ignore the result.

## Loading an XML Configuration File

In order to simplify the use and configuration of this script, we will be using eXtensible markup language (XML) as the document format for the script configuration file. Using XML, we can define the threshold information that we need to work with and avoid hard-coding values into the threshold script itself. The XML document shown in Figure III.11 shows one way of storing the configuration values we'll be working with in the XML format.

**Figure III.11** Threshold Configuration XML

```xml
<threshold>
   <monitor name="sys_proc">
      <status name="red" value="90" operator="equalorgreater">
         <action>email</action>
         <action>page</action>
      </status>
      <status name="yellow" value="85"
operator="equalorgreater">
         <action>email</action>
      </status>
      <status name="green" value="85" operator="less">
         <action>none</action>
      </status>
    </monitor>
   <monitor name="sys_mem">
      <status name="red" value="50" operator="equalorless">
         <action>email</action>
         <action>page</action>
      </status>
      <status name="yellow" value="100" operator="equalorless">
         <action>email</action>
      </status>
      <status name="green" value="100" operator="greater">
         <action>none</action>
      </status>
   </monitor>
</threshold>
```

In this XML document, we have set up a hierarchical format for the configuration information. The first tag shown in the document is <threshold>, which indicates what the contained values will be used for. We then use the <monitor> tag to indicate a specific monitor and set a "name" value for the monitor. This is followed by a series of <status> tags, each with their own "name", "value," and "operator" parameters. Within the <status> tags, one or more <action> tags exist indicating which actions should be taken when the conditions set for the status are met.

To import this XML data and make use of it, we will be using the XML::Simple module. This module is loaded by default with ActivePerl, but can be installed using PPM if your ActivePerl installation does not have it installed. The XML::Simple module allows for the importing, exporting, and manipulation of XML data from a file or string. To use our XML file, one of the first actions we will take within our threshold script will be to import the XML using the XML::Simple module. An easy way to perform this action is shown in Figure III.12.

### Figure III.12 load_display_XML.pl

```perl
use strict;
use XML::Simple;
use Data::Dumper;

my $xml = new XML::Simple(KeyAttr=>[], ForceArray => 1);
my $data=$xml->XMLin("/dls/thresh.xml");

#DEBUG
#Uncomment to print data structure
#print Dumper($data);
#END DEBUG

foreach my $monitor (@{$data->{monitor}})
{
    print "Monitor: " . $monitor->{name} . "\n";
    foreach my $status (@{$monitor->{status}})
    {
        foreach my $action (@{$status->{action}})
        {
            print "If value is " . $status->{operator} .
                " than " . $status->{value}. " take action " .
                $action . " and set status as " .
                $status->{name} . ".\n";
        }
    }
}
print "\n\nFinished!\n";
```

When we run the code shown in Figure III.12, we obtain the results shown in Figure III.13.

**Figure III.13** load_display_XML Results

Basically this is an output of all the rules built into our XML file. Let's walk through the code. First, we have our "use" statements indicating that we will be using XML::Simple and Data::Dumper. While Data::Dumper is not necessary for parsing the XML, this module can be very useful in displaying a data structure. You'll see the use of this a little farther down in the code in the "DEBUG" section.

Next, we create a new instance of XML::Simple with a couple of options. The first option is the "KeyAttr" option, which translates nested elements from an array to a hash. We're also using the "ForceArray" option to force XML::Simple to create arrays even if there is only one element. This makes it a lot easier to write code to consistently handle the data in the XML. For a full description of the options available for XML::Simple, please see http://search.cpan.org/~grantm/XML-Simple-2.18/lib/XML/Simple.pm.

The following line of code loads in our XML file using the "XMLin" function. Then we have our debug section. If you uncomment the code calling the "Dumper" function, the script will print out the complete data structure as it is imported from the XML file. This can be very helpful in writing code to handle the structure, as you can visually see the data elements that you are working with.

Now we can start walking through the data elements. We start by using the monitors indicated in the XML file. As we go through each data element, we need to keep in mind that we are dealing with arrays of hashes for the most part. Each element of our XML contains a structure underneath the element with the exception of the "action" element, which just contains a simple array. Once we populate the $monitor hash with the data stored in $data->{monitor}, we can easily work with the data elements within the hash. This is shown in the next line where we print the name of the specific monitor that we are looking at.

After printing the monitor name, we then need to display the status and action information. We do this by looping through the hash elements in $monitor->{status} and $status->{action}, respectively. Again note that $action is loaded as an array of scalar values, not hashes. Therefore, in our print statement, we refer to each hash value by name, but just display the action using $action. The end result of this very long print statement is to show the actions associated with each status for each monitor.

This is useful code for displaying the data shown in our XML and demonstrating how that data can be used, but it really doesn't do very much. It's just intended as a listing mechanism and an example of how we can use XML to control how our thresholds work. So let's make this its own subroutine called "threshold_rules" and include it in our overall threshold management script for future reference.

## Evaluating Thresholds

Next we need to write a script that will take the XML we have imported and evaluate it against the performance indicator data we gather from our monitoring scripts. To do this, we'll create another subroutine called "threshold_check" to handle evaluating data. We'll also need to handle evaluation of actions, so we'll use a subroutine called "take_action" for that purpose. The script shown in Figure III.14 shows an example of how this can be done.

## Figure III.14 thresh.pl

```perl
sub threshold_rules {
    use XML::Simple;
    use Data::Dumper;

    my $xml = new XML::Simple(KeyAttr=>[], ForceArray => 1);
    my $data=$xml->XMLin("/dls/thresh.xml");
    #Uncomment to print data structure
    #print Dumper($data);
    foreach my $monitor (@{$data->{monitor}})
    {
        print "Monitor: " . $monitor->{name} . "\n";
        foreach my $status (@{$monitor->{status}})
        {
            foreach my $action (@{$status->{action}})
            {
                print "If value is " . $status->{operator} .
                    " than " . $status->{value}.
                    " take action " . $action .
                    " and set status as " .
                    $status->{name} . ".\n";
            }
        }
    }
    print "\n\nFinished!\n";
}

sub threshold_check {
use XML::Simple;
use Data::Dumper;

    my $xml = new XML::Simple(KeyAttr=>[], ForceArray => 1);
    my $data=$xml->XMLin("/dls/thresh.xml");
    foreach my $monitor (@{$data->{monitor}})
    {
        if ($monitor->{name} eq $_[0])
        {
            print "Match found for " . $monitor->{name} .
                "!\n";

            foreach my $status (@{$monitor->{status}})
            {
                foreach my $action (@{$status->{action}})
                {
                    print "Evaluating rule \"If value is " .
                        $status->{operator} . " than " .
                        $status->{value}. " perform action " .
                        $action . " and set status as " .
                        $status->{name} . "\".\n";

                    if ($status->{operator} eq
                        "equalorgreater") {
                        if ($_[1] >= $status->{value}) {
                            &take_action($action,
                                        $status->{name});
                        }
```

```perl
                         } elsif ($status->{operator} eq
                                  "equalorless") {
                             if ($_[1] <= $status->{value}){
                                 &take_action($action,
                                              $status->{name});
                             }
                         } elsif ($status->{operator} eq
                                  "equal") {
                             if ($_[1] = $status->{value}){
                                 &take_action($action,
                                              $status->{name});
                             }
                         } elsif ($status->{operator} eq
                                  "notequal") {
                             if ($_[1] != $status->{value}){
                                 &take_action($action,
                                              $status->{name});
                             }
                         } elsif ($status->{operator} eq
                                  "greater") {
                             if ($_[1] > $status->{value}){
                                 &take_action($action,
                                              $status->{name});
                             }
                         } elsif ($status->{operator} eq
                                  "less") {
                             if ($_[1] < $status->{value}){
                                 &take_action($action,
                                              $status->{name});
                             }
                         }
                     }
                 }
             }
         }
}

sub take_action
{
    if ($_[0] eq "none") {
        print "Taking no action";
    } else {
        print "Taking action!!!  We need to $_[0] someone!\n" .
            "Condition is $_[1]!\n";
    }
}

1;
```

Now we're developing a fairly long script containing a variety of subroutines for dealing with threshold-related items. Again, in the interest of modularization, it makes sense to group these together and just call them from another script.

We've already walked through the code contained in the "threshold_rules" subroutine, so let's take a look at the "threshold_check" subroutine. We start off loading the XML data in the same manner used in "threshold_check". Then we do a string match against the monitors listed in the XML file to see if the first incoming variable ($_[0]) matches a known monitor. If it does, we gather the value and operator for the variety of statuses and actions associated to the monitor. For debugging purposes, we then print out which rule we're evaluating before moving on to the actual evaluations.

We have a set of operators available to check values. They are:

- equalorgreater
- equalorless
- equal
- notequal
- greater
- less

Using these operators, we can check the value variable ($_[1]) against the threshold value ($status->{value}). The series of if/elsif statements perform this evaluation. Note that we are specifically comparing numeric values here. Consequently, we have to make sure that we are dealing with a numeric value for that second variable. We'll need to put in some error checking for this and will demonstrate this in the next version of the script.

When a match is found, the subroutine "take_action" is called passing a variable containing the action name and status code. Right now we're just printing that string, but we can do more with that later. In this area, we're evaluating all possible rules for the monitor against the value. That means that we could potentially have more than one match. For example, using the following test script:

```
use strict;
require 'thresh.pl';
&threshold_check ("sys_proc" , 85);
```

yields the results shown in Figure III.15, which has a single match for a yellow status code.

**Figure III.15** Threshold Test Results

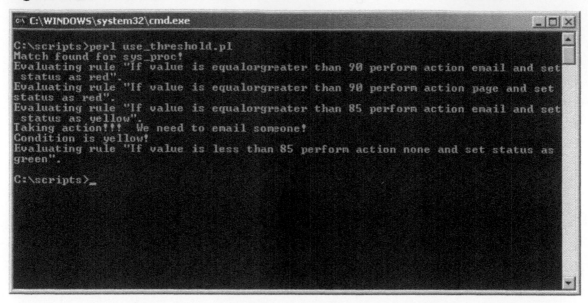

However, if we change the threshold value in our test script to 90, we end up with three matches; two red status actions and one yellow. This is shown in Figure III.16.

**Figure III.16** Threshold Test Results 2

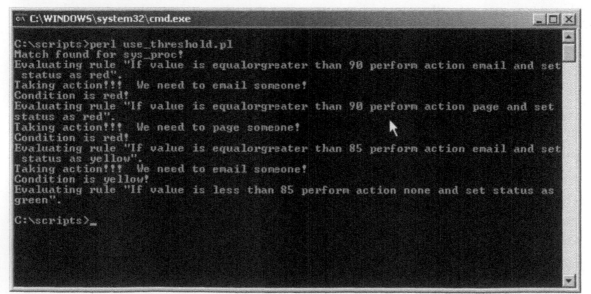

Make sure that you keep this behavior in mind when using these scripts for monitoring. If a status condition is moving from green to yellow to red, you will receive actions for all of the conditions if they overlap. If you need to handle these conditions separately, you can modify the evaluation logic by either stopping the evaluation when a match is found working in a specific order (so that it doesn't stop evaluation at a match in yellow when it would have matched a red condition), or setting a condition variable and updating it as the condition becomes progressively worse (i.e., update the variable from yellow to red, but not from red to yellow). For the purposes of our examples, multiple matches are acceptable.

One additional thing to note prior to moving on to examining the "take_action" subroutine in detail, is the last line of the code sample. The "1;" line is critical to ensuring that this code functions properly. Since we're including the thresh.pl file in another Perl script, the "use" statement must evaluate to true when the thresh.pl script is loaded. To accomplish this, we add "1;" as the last line of the script.

## Taking Action

Again, our "take_action" subroutine is very simple and just outputs the two variables that are passed to it. This is where we can have some fun and define all of our possible actions for various conditions! You can do anything you want here such as call a script to automatically put in a ticket in your helpdesk system, send a Simple Network Management Protocol (SNMP) trap, or even page an application administrator. For our example, we'll just set up an e-mail action and a page action using an e-mail-based paging device. This will basically be the same e-mail functionality, but the messages will have to be formed differently due to text space limitations for the pager.

Figure III.17 shows how our "thresh.pl" script can be expanded to include some error checking around the incoming data and the two new actions for e-mailing and paging.

## Figure III.17 thresh.pl

```perl
require 'actions.pl';

sub threshold_rules {
    use XML::Simple;
    use Data::Dumper;

    my $xml = new XML::Simple(KeyAttr=>[], ForceArray => 1);
    my $data=$xml->XMLin("/dls/thresh.xml");
    #Uncomment to print data structure
    #print Dumper($data);
    foreach my $monitor (@{$data->{monitor}})
    {
        print "Monitor: " . $monitor->{name} . "\n";
        foreach my $status (@{$monitor->{status}})
        {
            foreach my $action (@{$status->{action}})
            {
                print "If value is " . $status->{operator} .
                    " than " . $status->{value}.
                    " take action " . $action .
                    " and set status as " .
                    $status->{name} . ".\n";
            }
        }
    }
    print "\n\nFinished!\n";
}

sub threshold_check {
use XML::Simple;
use Data::Dumper;

if(defined($_[0]) && defined($_[1]) && int($_[1])) {
    my $xml = new XML::Simple(KeyAttr=>[], ForceArray => 1);
    my $data=$xml->XMLin("/dls/thresh.xml");
    foreach my $monitor (@{$data->{monitor}})
    {
        if ($monitor->{name} eq $_[0])
        {
            print "Match found for " . $monitor->{name} .
                "!\n";

            foreach my $status (@{$monitor->{status}})
            {
                foreach my $action (@{$status->{action}})
                {
                    print my $rule="Evaluating rule \"If " .
                        "value is " .
                        $status->{operator} . " than " .
                        $status->{value}. " perform action " .
                        $action . " and set status as " .
                        $status->{name} . "\".\n";

                    if ($status->{operator} eq
                        "equalorgreater") {
                        if ($_[1] >= $status->{value}) {
                            &take_action($action,
```

```perl
                                $status->{name},
                                    $rule);
                }
            } elsif ($status->{operator} eq
                    "equalorless") {
                if ($_[1] <= $status->{value}){
                    &take_action($action,
                                $status->{name},
                                    $rule);
                }
            } elsif ($status->{operator} eq
                    "equal") {
                if ($_[1] == $status->{value}){
                    &take_action($action,
                                $status->{name},
                                    $rule);
                }
            } elsif ($status->{operator} eq
                    "notequal") {
                if ($_[1] != $status->{value}){
                    &take_action($action,
                                $status->{name},
                                    $rule);
                }
            } elsif ($status->{operator} eq
                    "greater") {
                if ($_[1] > $status->{value}){
                    &take_action($action,
                                $status->{name},
                                    $rule);
                }
            } elsif ($status->{operator} eq
                    "less") {
                if ($_[1] < $status->{value}){
                    &take_action($action,
                                $status->{name},
                                    $rule);
                }
            }
        }
    }
} else {
    print "Input values do not meet requirements!\n";
}
}

sub take_action {
    if(defined($_[0]) && defined($_[1]) && defined($_[2])) {
    if ($_[0] eq "none") {
        print "Taking no action";
    } elsif ($_[0] eq "email") {
        &action_email("admin\@example.com",
                    "Code $_[1] alert!",
                    "Rule violated: $_[2]");
```

```
        } elsif ($_[0] eq "page") {
            &action_page("admin_pager\@example.com",
                         "Code $_[1]",
                         "$_[2]");
        } else {
            print "Invalid Action!";
        }
    } else {
        print "Input values do not meet requirements!\n";
    }
}

1;
```

There are a few changes in the script that should be noted in this revision. First, we have added a new dependency file called "actions.pl" and included a "require" statement to load in this file. Next, we've added some quick error checking code using the following line:

```
if(defined($_[0]) && defined($_[1]) && int($_[1])) {
```

This basically checks to ensure that we have two values being passed to the subroutine ($_[0] and $_[1]) and that the second value is an integer. If any of these three requirements are not met, we print an error message stating that the "Input values do not meet requirements." This is a very basic error check and certainly should be expanded to cover a variety of contingencies. Just like any other code sample in a book, you should modify these scripts to fit your needs and handle error conditions that are specific to your environment.

There are two other changes in the "threshold_check" subroutine that you should pay attention to. First is the modification of the print statement used to display the current rule being evaluated. This string is now being stored in the $rule variable for use with our actions. Secondly, we have changed the calls to the "take_action" subroutine to include another required variable; the rule being evaluated.

The "take_action" subroutine has also seen some changes. First, we've added some more error checking to ensure that we have three incoming variables. Next, we're now using a if/elsif/else evaluation of the first incoming variable ($_[0]) to determine which action routine we should take. Lastly, we're calling new subroutines for e-mailing and paging that are coming out of the actions.pl file.

Let's take a look at the actions.pl file and see what the two subroutines we're calling ("action_email" and "action_page") actually do. This script is shown in Figure III.18.

**Figure III.18** actions.pl

```perl
sub action_email {
if(defined($_[0]) && defined($_[1]) && defined($_[2])) {
use Mail::Sender;
    $sender = new Mail::Sender {
            smtp => 'localhost',
            from => 'admin@example.com',
            on_errors => undef,
    }
            or die "Can't create the Mail::Sender object: " .
                "$Mail::Sender::Error\n";
    $sender->Open({
            to => $_[0],
            subject => $_[1]
    })
            or die "Can't open the message: ".
                "$sender->{'error_msg'}\n";
    $sender->SendLineEnc($_[2]);
    $sender->Close()
            or die "Failed to send the message: " .
                "$sender->{'error_msg'}\n";
}
}

sub action_page {
if(defined($_[0]) && defined($_[1]) && defined($_[2])) {
use Mail::Sender;
    $sender = new Mail::Sender {
            smtp => 'localhost',
            from => 'admin_pager@example.com',
            on_errors => undef,
    }
            or die "Can't create the Mail::Sender object: " .
                "$Mail::Sender::Error\n";
    $sender->Open({
            to => $_[0],
            subject => $_[1]
    })
            or die "Can't open the message: ".
                "$sender->{'error_msg'}\n";
    $sender->SendLineEnc($_[2]);
    $sender->Close()
            or die "Failed to send the message: " .
                "$sender->{'error_msg'}\n";
}
}

1;
```

In Figure III.18 we have defined the two subroutines for "action_email" and "action_page." For example purposes, they are nearly identical with exception for the "from" e-mail header. In reality, it would be more efficient to use the same subroutine for this, but since some paging providers have unique requirements, I have separated the two subroutines so that they can be easily modified independently.

First, we do some very basic error checking to make sure that we have the correct number of options being sent to the subroutine. Again, expand this error checking to include all necessary checks for your own environment. A good addition would be using regular expressions to confirm that the e-mail address is in the right format.

With that in place, we then add in the use of a new module. The Mail::Sender module is not included in the base ActivePerl installation, but is very useful for handling e-mail-related needs. For a full listing of its options, please see http://search. cpan.org/dist/Mail-Sender-0.8.13/Sender.pm. This module will consequently do all of our e-mail handling for us so we don't have to worry about opening ports manually, creating an Simple Mail Transfer Protocol (SMTP) message in the correct format, and so forth.

Next, we create a new Mail::Sender object with a variety of options to set our SMTP mail server address, "from" header, and the error handling option. Then we call the "Open" function of the object and pass in our incoming variables for the "to" and subject headers. Finally, we send the message using the "SendLineEnc" function. Again, all of this is duplicated in the "action_page" subroutine with exception of the "from" header.

## Putting it all Together

By calling our new subroutine in the manner shown in Figure III.17, we are now able to send e-mails and pages to our application admin from our test script. The last thing we need to do to make all of this functional is to get away from using the test script and actually integrating the threshold management script and action script with our monitoring scripts. Figure III.19 and III.20 show final versions of the get_system_stats.pl and thresh.xml files.

## Figure III.19 get_system_stats.pl

```perl
#get_system_stats.pl
use Win32::PerfMon;
use strict;
require 'thresh.pl';

my $ret = undef;
my $err = undef;
my $Object = undef;
my $Counter = undef;
my $CounterData = undef;
#connect to localhost for data
my $perfmon = Win32::PerfMon->new("\\\\localhost");

if($perfmon != undef)
{
    $ret = $perfmon->AddCounter("Processor",
                "% Processor Time", "_Total");
    if($ret != 0) {
        $ret = $perfmon->AddCounter("Memory",
                "Available MBytes", -1);
    }
    if($ret != 0) {
        $ret = $perfmon->AddCounter("Network Interface",
                "Bytes Total/sec",
                "Intel 21140-Based PCI Fast ".
                "Ethernet Adapter [Generic]");
    }
    if($ret != 0) {
        $ret = $perfmon->AddCounter("Network Interface",
                "Current Bandwidth",
                "Intel 21140-Based PCI Fast ".
                "Ethernet Adapter [Generic]");
    }
    if($ret != 0)
    {
        $ret = $perfmon->CollectData();
        if($ret != 0)                    {
            my $proctime = $perfmon->GetCounterValue(
                "Processor","% Processor Time",
                "_Total");
            if($proctime > -1)
            {
            print "% Processor Time = [$proctime]\n";
            &threshold_check ("sys_proc", $proctime);
            }
            else
            {
            $err = $perfmon->GetErrorText();
            print "Failed to get the counter data!\n",
                $err, "\n";
            }

            my $freemem = $perfmon->GetCounterValue(
                "Memory","Available MBytes",
                -1);
              if($freemem > -1)
```

```
{
print "Available Memory = [$freemem]MB\n";
&threshold_check ("sys_mem", $freemem);
}
else
{
$err = $perfmon->GetErrorText();
print "Failed to get the counter data!\n",
      $err, "\n";
}

my $netbps = $perfmon->GetCounterValue(
      "Network Interface",
      "Bytes Total/sec",
      "Intel 21140-Based PCI Fast ".
      "Ethernet Adapter [Generic]");
  if($netbps > -1)
{
print "Network Bytes/Second = [$netbps]\n";
if ($netbps > 0) {
&threshold_check ("sys_netbps",
                   int($netbps));
}
}
else
{
$err = $perfmon->GetErrorText();
print "Failed to get the counter data!\n",
      $err, "\n";
}

my $netbw = $perfmon->GetCounterValue(
      "Network Interface",
      "Current Bandwidth",
      "Intel 21140-Based PCI Fast ".
      "Ethernet Adapter [Generic]");
  if($netbw > -1)
{
print "Network Bandwidth Bits/Second =".
      " [$netbw]\n";
}
else
{
err = $perfmon->GetErrorText();
print "Failed to get the counter data!\n",
      $err, "\n";
}

if ($netbps > -1 && $netbw > -1)
{
      my $netutil = sprintf "%.2f",
            (8 * $netbps / $netbw);
      print "% Network Utilization = ".
            " [$netutil]\n";
```

```
                        if ($netutil >0) {
                        &threshold_check ("sys_netutil",
                                            $netutil);
                        }
                }
                else
                {
                        print "Not enough data to determine".
                            "network utilization!";
                }
            }
            else
            {
                    $err = $perfmon->GetErrorText();
                    print "Failed to collect the perf data!\n",
                        $err, "\n";
            }
        }
        else
        {
            $err = $perfmon->GetErrorText();
            print "Failed to add the counter!\n", $err, "\n";
        }
    }
    else
    {
        print "Failed to create the perf object!\n";
    }
```

## Figure III.20 thresh.xml

```xml
<threshold>
    <monitor name="sys_proc">
        <status name="red" value="90" operator="equalorgreater">
            <action>email</action>
            <action>page</action>
        </status>
        <status name="yellow" value="85"
operator="equalorgreater">
            <action>email</action>
        </status>
        <status name="green" value="85" operator="less">
            <action>none</action>
        </status>
     </monitor>
    <monitor name="sys_mem">
        <status name="red" value="50" operator="equalorless">
            <action>email</action>
            <action>page</action>
        </status>
        <status name="yellow" value="80" operator="equalorless">
            <action>email</action>
        </status>
        <status name="green" value="80" operator="greater">
            <action>none</action>
        </status>
     </monitor>
    <monitor name="sys_netbps">
        <status name="red" value="500" operator="equalorgreater"
```

```
        <action>email</action>
            <action>page</action>
        </status>
        <status name="yellow" value="300"
operator="equalorgreater">
            <action>email</action>
        </status>
        <status name="green" value="300" operator="less">
            <action>none</action>
        </status>
      </monitor>
    <monitor name="sys_netutil">
        <status name="red" value="90" operator="equalorgreater">
            <action>email</action>
            <action>page</action>
        </status>
        <status name="yellow" value="75"
operator="equalorgreater">
            <action>email</action>
        </status>
        <status name="green" value="75" operator="less">
            <action>none</action>
        </status>
      </monitor>
</threshold>
```

The thresh.xml file shown in Figure III.20 is pretty straightforward. We have simply added in additional monitors for "sys_netbps" and "sys_netutil." Watch for the line wrapping in the printed XML! Additionally, the values shown in this example file are just samples. Thresholds should be configured to appropriate values for your environment. To extend this file to handle thresholds for the get_process_stats.pl, we would need to add in more monitors and configure them.

The get_system_stats.pl in Figure III.19 has had a few changes. First, and most important, note the new "require" statement at the beginning used to load in the thresh.pl file. The second change is the addition of lines similar to:

```
&threshold_check ("sys_proc", $proctime);
```

to each monitoring area. Basically, we are identifying the monitor to use and passing the statistical data to the "threshold_check" subroutine. In some cases, we have to handle special values a little differently. For example, in the section where we are getting the Network Bytes/Second, there is the possibility that the value could be 0. This would be rejected by the "threshold_check" subroutine as it is not technically an integer. So, we handle this by not calling the threshold script if a 0 is found. Keep in mind that this may not be the behavior that you want in some cases. For example, if a 0 being returned from a monitor means that there is a problem, you may need to send an alert. In a case like this, just change the 0 to another value prior to sending it to the "threshold_check" subroutine, and use the values specified in your XML file to handle the new value.

When running the script, you should receive output similar to that shown in Figure III.21.

**Figure III.21** Get System Status Result

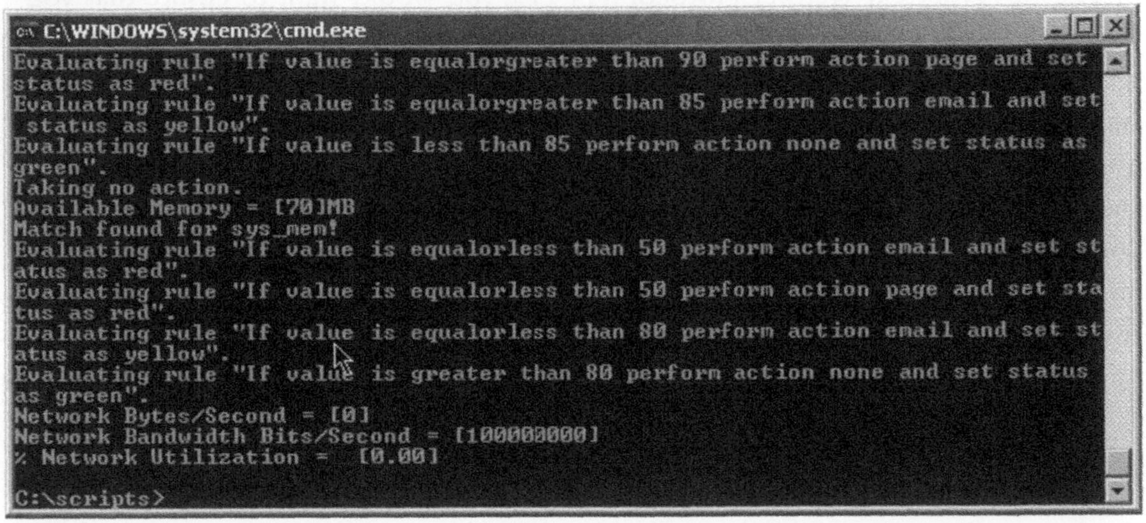

Now we have a collection of Perl scripts that monitor a core application. Through the collection of system and process statistics, a threshold evaluation script, and an action script all configured by an XML file, we are able to accomplish our goals of gathering critical data, evaluating that data, and taking action based on the results. Through simple modifications of these scripts, you can put a robust monitoring system in place for your core application.

# Core Application Dependencies

With monitoring in place for the core application, we can move on to building out some monitoring tools for core application dependencies. These dependencies can range from remote databases to SAN systems and even to other application processes. In order to have a good overview of what is happening in our application environment, these other factors and dependencies need to be monitored. In this section, we will build out some tools to monitor these application dependencies.

# Monitoring Remote System Availability

When monitoring a remote system, there are several factors to watch for. First, you want to make sure that the system is available on your network. By performing a simple PING check, this can easily be confirmed. But that does not necessarily mean that the system is actually available and providing the services that you need. For example, if you are relying on the system to respond on a specific port for a service such as SMTP or Hypertext Transfer Protocol (HTTP), you would want to ensure that you are able to communicate between systems on that port as well.

Doing this type of availability check is a little more complex, but certainly provides more information to us from a monitoring perspective. The easiest type of check for availability along these lines would be to open a port on the remote system, confirm that the communication does successfully take place, and then close the port. A more complex check would involve sending specific data down that communication channel and confirming the results, but at this point it would be wiser to build a reusable monitoring component that can check one or more ports for us.

The script shown in Figure III.22 shows one way that we can accomplish this type of remote system availability check using the IO::Socket module. This module is part of the default installation for ActivePerl.

**Figure III.22** get_port_status.pl

```
#get_port_status.pl
use strict;
require 'thresh.pl';
use IO::Socket;

my $portCheck = new IO::Socket::INET (
    PeerAddr => 'localhost',
    PeerPort => '80',
    Proto => 'tcp',
    );

if ($portCheck == undef) {
    &threshold_check ("remote_port", 1);
    } else {
    &threshold_check ("remote_port", 100);
    }

if ($portCheck != undef ) {
    close($portCheck);
}
```

This also requires a modification to our "thresh.xml" file in order to create a threshold for the new monitor. The additional XML for this monitor is shown in Figure III.23.

**Figure III.23** remote_port Monitor XML

```
<monitor name="remote_port">
    <status name="red" value="1" operator="equal">
       <action>email</action>
       <action>page</action>
    </status>
    <status name="green" value="100" operator="equal">
       <action>none</action>
    </status>
</monitor>
```

In this script we're doing a very simple check of the port status on the remote system. If this test is successful, we learn two things. First, that the remote system is available and secondly that it is accepting connection requests on the port we are checking. If this test is unsuccessful, we then know that either the remote system is down or the remote port is not responding.

The script itself begins by adding in the requirement of "thresh.pl" and the use of IO::Socket. Next, we create a new object utilizing the IO::Socket module with the address, port, and protocol used by the remote service that we are monitoring. From there it's just a matter of checking to see if we were successful in creating that connection or not and alerting as appropriate.

You'll note that for the status values we're using 100 and 1. Normally you would expect to see a 0 and 1 status code here, but remember that with the threshold validation script we require that the values being passed be integers. Since 0 is not an integer, we have to use other values for our status codes.

Lastly, and perhaps most importantly, we have to do some cleanup. If the "port Check" object is defined indicating a success, we must consequently close that port. You certainly don't want to open up a huge number of ports on your remote system due to your monitoring script. Using the "close" function we can accomplish this cleanup.

# Monitoring Available Disk Space

Another dependency that is often overlooked is disk space. When monitoring an application, it is pretty typical to forget that disk space is a requirement even for applications that are not disk intensive. There are always log file writes, movement of temporary data, and potential writes for virtual memory use to consider. With that in mind, we need a script to monitor free disk space.

While we could use the performance monitor statistics for this, there could be some potential pitfalls with that approach. Performance monitor only collects disk statistics if they are explicitly enabled through the use of a command line action. The alternative is to manually check the disk information. While this is slightly more disk intensive, it is more reliable and will provide useful information in more situations.

The script shown in Figure III.24 demonstrates how this type of free disk space check can be performed. This script makes use of the Win32::FileOp module, which in turn has dependencies on the Data::Lazy, Win32::AbsPath, and Win32::API modules. More information on the Win32::FileOp module can be found at http://search.cpan.org/dist/Win32-FileOp-0.14.1/FileOp.pm.

### Figure III.24 get_disk_space.pl

```
#get_disk_space.pl
use strict;
require 'thresh.pl';
use Win32::FileOp qw(GetDiskFreeSpace);

my $disk="c:";

(my $freeSpaceForUser, my $totalSize, my $totalFreeSpace)
    = GetDiskFreeSpace $disk;

if ($totalFreeSpace != undef and $totalSize != undef) {
    my $percentFreeSpace = int(100*($totalFreeSpace
        / $totalSize));
    &threshold_check ("free_disk_space", $percentFreeSpace);
    } else {
    print "Unable to get disk status!\n";
    }
```

And again there are some changes to our "thresh.xml" file to handle this additional monitor. The new monitor XML is shown in Figure III.25.

**Figure III.25** free_disk_space Monitor XML

```xml
<monitor name="free_disk_space">
  <status name="red" value="10" operator="equalorless">
    <action>email</action>
    <action>page</action>
  </status>
  <status name="yellow" value="20" operator="equalorless">
    <action>email</action>
  </status>
  <status name="green" value="20" operator="greater">
    <action>none</action>
  </status>
</monitor>
```

This simple script quickly gathers the disk information that we need and provides it to the "thresh.pl" script for handling. First, we specify the use of the Win32::FileOp module and specifically the "GetDiskFreeSpace" function. This function pulls the free disk space for the user, the total disk space, and the total free disk space from the drive specified in the $disk variable. We'll store all of this data into variables with intuitive names for later use.

After making sure that we were actually able to obtain the data we need through a definition check of the appropriate variables, we can move on to doing a quick percentage computation. By dividing the total free space by the total disk space and multiplying by 100, we are able to get the percentage of free space. This is a much easier number to work with than specifying the number of free bytes in the "thresh.xml" file. In order to use this value, we need to drop any data after the decimal point and can do this by running it through the "int" function to convert the value to an integer.

Lastly, we just call the "threshold_check" function and specify the monitor and value. This takes the data we added to the "thresh.xml" file and handles the actions appropriate to the thresholds specified.

# Monitoring Remote Disk Availability

In some cases you may need to ensure that a remote share is both available and has sufficient free space available for your uses. We can write a quick script for performing this function as well using the Win32::FileOp module. Since this need is similar to the need to gather local disk information, we can combine these scripts into a single script for the purpose of validating disk and free space availability. Figure III.26 shows how this can be accomplished.

**Figure III.26** get_disk_info.pl

```
#get_disk_info.pl
use strict;
require 'thresh.pl';
use Win32::FileOp qw(GetDiskFreeSpace);

my $disk="c:";
my $remote_disk="z:";
my $remote_share="\\\\localhost\\scripts";
my $freeSpaceForUser, my $totalSize, my $totalFreeSpace, my
$percentFreeSpace;

($freeSpaceForUser, $totalSize, $totalFreeSpace)
    = GetDiskFreeSpace $disk;
if ($totalFreeSpace != undef and $totalSize != undef) {
    $percentFreeSpace = int(100*($totalFreeSpace
        / $totalSize));
    &threshold_check ("free_disk_space", $percentFreeSpace);
    } else {
    print "Unable to get disk status!\n";
    &threshold_check ("free_disk_space", 1),
    }

($freeSpaceForUser, $totalSize, $totalFreeSpace)
    = GetDiskFreeSpace $remote_disk;
if ($totalFreeSpace != undef and $totalSize != undef) {
    $percentFreeSpace = int(100*($totalFreeSpace
        / $totalSize));
    &threshold_check ("free_disk_space", $percentFreeSpace);
    } else {
    print "Unable to get disk status!\n";
    &threshold_check ("free_disk_space", 1),
    }

($freeSpaceForUser, $totalSize, $totalFreeSpace)
    = GetDiskFreeSpace $remote_share;
if ($totalFreeSpace != undef and $totalSize != undef) {
    $percentFreeSpace = int(100*($totalFreeSpace
        / $totalSize));
    &threshold_check ("free_disk_space", $percentFreeSpace);
    } else {
    print "Unable to get disk status!\n";
    &threshold_check ("free_disk_space", 1),
    }
```

While this is similar to the code shown in Figure III.24, there are a few notable differences. First, we're declaring all of our variables up front since we will be reusing them multiple times within the script. Secondly, we have added a call to the "threshold_check" function passing a hard-coded value of 1 when we are unable to get the disk free space value. This has the effect of sending a value of 1 if the remote disk is not available. We simply have to make sure that we have an alert specified if a threshold value of 1 is received, which is already defined due to the values used in the XML shown in Figure III.25.

You'll also note that for all of these checks we are using different values for which disk we are checking, but using the same monitor for the threshold check. Typically, the same values will apply for a remote disk as they do for a local disk. If this is not the case, you can define a new monitor specific to the disks you are checking.

In this script we are effectively checking three disks: the local disk, a remote disk that has been mapped to a local drive, and a remote disk accessed by its share name. These are the most common types of disk access methods that you would use when connecting to a disk, and are all easily handled by the Win32::FileOp module.

## Monitoring Remote Databases

When an application reaches the enterprise level, it frequently separates the presentation layer of the application from the data layer. This means that a database is brought into the equation, whether local or remote. In order to monitor all dependencies of the application, it then becomes critical to monitor the database used by the application.

We could simply reuse the "get_port_status.pl" script shown in Figure III.22 for this purpose, but experience shows that while a connection may be able to be made to the listening port for a database, that does not necessarily mean that the database is available. This requires a more complex test than a simple port connection.

Rather than checking the port status, we need to go further and actually connect, authenticate, and confirm that the database is returning a valid response. For this example, we will be using the DBI module. This module is included in the ActivePerl installation and extensive detail on this module can be found at http://search.cpan. org/dist/DBI-1.58/

While there are other database connection modules available, this one allows for a great deal of flexibility and works well for monitoring purposes. There are also many drivers available that work with this module to allow for connections to a variety of database types. In this example, we'll be using the DBD::mysql module to connect to a local MySQL database. Details on this specific driver can be found at http://search. cpan.org/dist/DBD-mysql-3.0002/ and the module can be installed through PPM.

The script shown in Figure III.27 shows a simple database availability monitoring process that goes through all of the steps described above, and sends the results over to the usual threshold script. Additionally, the XML shown in Figure III.28 shows the necessary addition to the "thresh.xml" file to add these monitors.

**Figure III.27** get_db_avail.pl

```
#get_db_avail.pl
use strict;
require 'thresh.pl';
use MySQL;
my $host="localhost";
my $database="new_db";
my $user="test_user";
my $password="test12";

my $DBHandle = Mysql->connect($host, $database, $user,
$password);

if ($DBHandle == undef) {
    &threshold_check ("DB_avail", 1);
    } else {
    &threshold_check ("DB_avail", 100);
    }

if ($DBHandle != undef ) {
    undef $DBHandle;
    }
```

**Figure III.28** DB_avail Monitor XML

```
<monitor name="DB_avail">
    <status name="red" value="1" operator="equal">
        <action>email</action>
        <action>page</action>
    </status>
    <status name="green" value="100" operator="equal">
        <action>none</action>
    </status>
</monitor>
```

In effect, we are making a connection to the MySQL database using the parameters specified for $host, $database, $user, and $password. This database connection object is then stored in $DBHandle. Similar to the script shown in Figure III.22 for a port check, we then check and see if the $DBHandle variable is defined. If it is, we successfully made a connection and report a "1" to the "threshold_check" function. Otherwise, we report a "100" to indicate a failure. Then we do some cleanup and undef our connection to the database. The monitor used is detailed in the XML shown in Figure III.28.

# Monitoring Other Dependencies

Now that we have a few scripts in place for monitoring some application dependencies, we have a framework built for monitoring a great deal of other dependencies. Simply reuse these scripts as needed and convert them to the purpose you need. For example, let's say that you want to check to ensure that a subprocess you need is functioning. For this, you'd just reuse some of the code used in the "get_system_stats.pl" script shown in Figure III.19. Need to check to make sure that your system isn't paging too much? That's just another performance monitor counter and can be monitored using the same script.

The purpose of developing the framework for these scripts in the manner done in this section is to provide a great deal of flexibility. By modifying these scripts to fit your specific needs, you can quickly and easily build a robust monitoring system for your applications. By making small modifications to the scripts, they can be changed to monitor exactly what you need for your specific application.

Of course, there are some application monitoring needs that do require a little more complexity. The remainder of this chapter deals with those complex monitoring needs and will help you in expanding this basic monitoring into a full monitoring solution.

# Web Services

Web services and Service Oriented Architecture (SOA) are becoming more common as a method of interfacing multiple applications. This adds another layer of complexity to application monitoring as you have to rely on the "black box" of a Web service. Typically, these Web services do not provide a great deal of information about their internal operations. There are a few different monitors that can be built and configured to help in ensuring the availability of remote Web services, and we will be discussing a few of them here.

Web services are basically Web-facing interfaces for applications that use a standard document format (XML) for messages. Simple Object Access Protocol (SOAP) is the protocol used for transferring these XML messages between systems, and forms a standard method of access for interfaced systems.

## Monitoring Web Service Availability

The first step to monitoring Web services is to make sure that the service is available. Similar to monitoring for a database connection, we need to go further than a simple port check for validating Web service availability. There are a couple of ways that this can be accomplished. First would be to simply send a SOAP message to the Web service you are using and confirm that it responds as expected. The second is to send a SOAP message to a test Web service that is built specifically for validating availability of the system.

In this example, we will be using the SOAP::Lite module, which is included with ActivePerl or downloaded through PPM. The Web site for SOAP::Lite has a great deal of information on how to use the module as well as several test scripts. This information can be found at http://www.soaplite.com/.

Figure III.29 shows a script for testing the availability of a Web service using SOAP. This is a quick connectivity test, but also sends a valid SOAP Web service request as part of the test. The corresponding changes to the "thresh.xml" file to add the monitor for this are shown in Figure III.30.

## Figure III.29 get_ws_avail.pl

```
#get_ws_avail.pl
use SOAP::Lite;
use strict;
require 'thresh.pl';

my $soap = SOAP::Lite
  -> uri('http://www.soaplite.com/Temperatures')
  -> proxy('http://services.soaplite.com/temper.cgi');
my $result = $soap->f2c(61);
unless ($result->fault) {
  &threshold_check ("WS_avail", 100);
} else {
  &threshold_check ("WS_avail", 1);
}
```

## Figure III.30 WS_avail Monitor XML

```
<monitor name="WS_avail">
  <status name="red" value="1" operator="equal">
    <action>email</action>
    <action>page</action>
  </status>
  <status name="green" value="100" operator="equal">
    <action>none</action>
  </status>
</monitor>
```

This script is very similar to the database availability monitor script shown in Figure III.27. The main difference is that we are connecting to a data source which is a Web service using SOAP. The $soap object is created by passing parameters for the URI and proxy to SOAP::Lite. More information on how to configure these parameters can be found on the SOAP::Lite Web site at the URL mentioned above, but for our purposes the values in the script point to a temperature conversion Web service hosted by soaplite.com.

We then create a $result variable to hold the result of the Web service call and through the act of assigning this variable, we contact the Web service and call the "f2c" conversion function passing a value of 61. While this should return a specific value for the conversion, we're not concerned at this point with what that value is. Rather, we just need to confirm that a value is returned.

Handling the returned result provides the logic behind our monitoring. If the result is not a fault, we pass the value of 100 to the "threshold_check" function indicating that the Web service is responding. If a fault is encountered, however, we send a value of 1 and the "threshold_check" function takes the actions specified in our "WS_avail" monitor XML.

# Monitoring Web Service Functionality

More complex Web services may rely on a host of additional dependencies in order to return an expected result. For example, you may be connecting to an Oracle Web Service Manager Gateway, have your request forwarded to a Business Process Execution Language server, and then to a Web service, which pulls data from a remote database. This complex chain of transactions will eventually return a result to you, but it may not necessarily be what you expected.

Since any step in this chain could potentially generate an error, fully testing Web service functionality involves sending a request that goes through each step then finally returns a result that should be consistent for validation. If an error is encountered, it may be presented as a valid SOAP message but contain an error message rather than the result that you are looking for.

We'll make the assumption for testing purposes that the "f2c" conversion function that we used in the last test is actually a complex function that has to go through several steps in order to succeed. For example, and this is completely fictional, let's assume that it takes the value that we send, connects to a BPEL process to convert the number to a string, connects to a remote database to do a string match in order to find the corresponding Celsius value, runs that through a BPEL process to convert the string to a number, and then finally returns our result. Any failure in this chain of transactions could cause a failure, but would not necessarily generate a fault to SOAP::Lite.

## Master Craftsman

### End-to-end Web Service Testing

While the script shown here does a good job at determining whether or not a Web service and all its components are functioning, it does not provide a great deal of troubleshooting information. Just as we monitor multiple layers of our applications (system, network, database, and so forth), you should consider monitoring multiple layers of the Web service to speed up troubleshooting when something goes wrong.

A good way to do this without increasing the load on the system is to perform the validation test first. If, after running the validation test, the test fails, you can then run individual scripts that test each portion of the Web service (i.e., test the BPEL process manager server, the Web service gateway, and the back end data source). This will identify where in the stack the problem lies and allow you to get everything fixed more quickly.

With that in mind, the next script performs a similar transaction, but provides a guaranteed value as a result. This result is then validated and if the validation fails, we generate an alert. This script can be seen in Figure III.31 with the corresponding XML for the "thresh.xml" file in Figure III.32.

### Figure III.31 get_ws_status.pl

```perl
#get_ws_status.pl
use SOAP::Lite;
use strict;
require 'thresh.pl';

my $soap = SOAP::Lite
  -> uri('http://www.soaplite.com/Temperatures')
  -> proxy('http://services.soaplite.com/temper.cgi');
my $result = $soap->f2c(32);
unless ($result->fault) {
  if ($result->result() != 0) {
    &threshold_check ("WS_status", 1);
  } else {
    &threshold_check ("WS_status", 100);
  }
} else {
  &threshold_check ("WS_status", 1);
}
```

**Figure III.32** WS_status Monitor XML

```xml
<monitor name="WS_status">
    <status name="red" value="1" operator="equal">
        <action>email</action>
        <action>page</action>
    </status>
    <status name="green" value="100" operator="equal">
        <action>none</action>
    </status>
</monitor>
```

# Building a Monitoring System

Throughout this chapter we have put together a variety of scripts that can be used to monitor applications. In order to consider this a full monitoring package, however, we need to tie all the scripts together. To do this, we'll take the final versions of a number of these scripts, call them with the appropriate parameters, and use the appropriate dependency scripts for handling thresholds and alerting.

The scripts we'll be putting together for this are the following:

- get_system_stats.pl shown in Figure III.19

- get_process_stats.pl using a modified version shown in Figure III.33

- get_port_status.pl shown in Figure III.22

- get_disk_info.pl shown in Figure III.26

- get_db_avail.pl shown in Figure III.27

- get_ws_status.pl shown in Figure III.31

We'll also be using the dependency scripts "thresh.pl," "actions.pl," and the file "thresh.xml" as shown in Figure III.17, Figure III.18, and Figure III.34, respectively. First, please take a look at the modified "get_process_stats.pl" shown in Figure III.33 and the final "thresh.xml" shown in Figure III.34.

## Figure III.33 get_process_stats.pl

```perl
#get_process_stats.pl
use Win32::PerfMon;
use strict;
use Win32::Process::Info;
require 'thresh.pl';

my $pi = Win32::Process::Info->new();
my @procinfo = $pi->GetProcInfo();
my $procname = undef;
my $procstatus=0;
my $ret = undef;
my $err = undef;
my $Object = undef;
my $Counter = undef;
my $CounterData = undef;
#connect to localhost for data
my $perfmon = Win32::PerfMon->new("\\\\localhost");

my $result = open PIDFILE, "Komodo.pid";
if ($result) {
    if ( defined(my $piddata = <PIDFILE>)) {
    chomp $piddata;
    for my $pid (@procinfo){
        if ($pid->{"ProcessId"} == $piddata) {
            $procname = $pid->{"Name"};
            print $pid->{"Name"}." is running!\n";
            &threshold_check ("proc_avail", 100);
            $procstatus=1;
        }
    }

    if ($procstatus!=1)
    {
        print "Process $piddata cannot be found!";
        &threshold_check ("proc_avail", 1);
    } else {
        $procname =~ s/^(.+?)(\.[^.]*)?$/$1/;
        if($perfmon != undef) {
            $ret = $perfmon->AddCounter("Process",
                "% Processor Time", $procname);
            if($ret != 0) {
                $ret = $perfmon->AddCounter("Process",
                    "Private Bytes", $procname);
            }
            if($ret != 0) {
                $ret = $perfmon->CollectData();
                if($ret != 0) {
                my $proctime=$perfmon->GetCounterValue(
                    "Process","% Processor Time",
                    $procname);
                    if($proctime > -1) {
                    print "% Processor Time = ".
                        "[$proctime]\n";
                    if ($proctime > 0) {
                        &threshold_check("proc_cpu",
```

```
                                              $proctime);
                                         } else {
                                             &threshold_check("proc_cpu",
                                             1);
                                         }
                                     } else {
                                         $err = $perfmon->GetErrorText();
                                         print"Failed to get the ".
                                         "processor counter data!\n",
                                         $err, "\n";
                                     }
                         my $freemem = $perfmon->GetCounterValue(
                                         "Process","Private Bytes",
                                         "komodo");
                                         if($freemem > -1)
                                         {
                         &threshold_check ("proc_mem",
                                              $freemem);
                         $freemem =~ s/(?<=\d)(?=(?:\d\d\d)+\b)/,/g;
                         print "Memory used by process = [$freemem]" .
                                 " Bytes\n";
                                     } else {
                                         $err = $perfmon->GetErrorText();
                         print "Failed to get the memory ".
                                 "counter data!\n",
                                 $err, "\n";
                                         }
                                     } else {
                                         $err = $perfmon->GetErrorText();
                                         print "Failed to collect the ".
                                              "perf data!\n", $err, "\n";
                                     }
                         } else {
                                 $err = $perfmon->GetErrorText();
                                 print "Failed to add the counter!\n",
                                         $err, "\n";
                         }
                 } else {
                         print "Failed to create the perf object!\n";
                         }
                     }
         } else {
                 print "PID not found in PID file.";
         }
} else {
         print "PID file not found.";
}
```

## Figure III.34 thresh.xml

```xml
<threshold>
  <monitor name="sys_proc">
    <status name="red" value="90" operator="equalorgreater">
      <action>email</action>
      <action>page</action>
    </status>
    <status name="yellow" value="85"
operator="equalorgreater">
      <action>email</action>
    </status>
    <status name="green" value="85" operator="less">
      <action>none</action>
    </status>
  </monitor>
  <monitor name="sys_mem">
    <status name="red" value="50" operator="equalorless">
      <action>email</action>
      <action>page</action>
    </status>
    <status name="yellow" value="80" operator="equalorless">
      <action>email</action>
    </status>
    <status name="green" value="80" operator="greater">
      <action>none</action>
    </status>
  </monitor>
  <monitor name="sys_netbps">
    <status name="red" value="5000" operator="equalorgreater">
      <action>email</action>
      <action>page</action>
    </status>
    <status name="yellow" value="3000" operator="equalorgreater">
      <action>email</action>
    </status>
    <status name="green" value="3000" operator="less">
      <action>none</action>
    </status>
  </monitor>
  <monitor name="sys_netutil">
    <status name="red" value="90" operator="equalorgreater">
      <action>email</action>
      <action>page</action>
    </status>
    <status name="yellow" value="75" operator="equalorgreater">
      <action>email</action>
    </status>
    <status name="green" value="75" operator="less">
      <action>none</action>
    </status>
  </monitor>
  <monitor name="remote_port">
    <status name="red" value="1" operator="equal">
      <action>email</action>
      <action>page</action>
    </status>
```

```
            <status name="green" value="100" operator="equal">
                <action>none</action>
        </status>
        </monitor>
    <monitor name="free_disk_space">
        <status name="red" value="10" operator="equalorless">
                <action>email</action>
                <action>page</action>
        </status>
<status name="yellow" value="20" operator="equalorless">
                <action>email</action>
        </status>
        <status name="green" value="20" operator="greater">
                <action>none</action>
        </status>
        </monitor>
    <monitor name="DB_avail">
        <status name="red" value="1" operator="equal">
                <action>email</action>
                <action>page</action>
        </status>
        <status name="green" value="100" operator="equal">
                <action>none</action>
        </status>
        </monitor>
    <monitor name="WS_status">
        <status name="red" value="1" operator="equal">
                <action>email</action>
                <action>page</action>
        </status>
        <status name="green" value="100" operator="equal">
                <action>none</action>
        </status>
        </monitor>
    <monitor name="proc_avail">
        <status name="red" value="1" operator="equal">
                <action>email</action>
                <action>page</action>
        </status>
        <status name="green" value="100" operator="equal">
                <action>none</action>
        </status>
        </monitor>
    <monitor name="proc_cpu">
        <status name="red" value="90" operator="equalorgreater">
                <action>email</action>
                <action>page</action>
        </status>
        <status name="yellow" value="85"
operator="equalorgreater">
                <action>email</action>
        </status>
        <status name="green" value="85" operator="less">
                <action>none</action>
        </status>
        </monitor>
    <monitor name="proc_mem">
```

```
<status name="red" value="90000000"
operator="equalorgreater">
        <action>email</action>
        <action>page</action>
      </status>
      <status name="yellow" value="80000000"
operator="equalorgreater">
         <action>email</action>
      </status>
      <status name="green" value="80000000" operator="less">
         <action>none</action>
      </status>
    </monitor>
</threshold>
```

Make sure you are watching for the wrapping in Figure III.34, as the width of some lines is longer than the page width in this book.

The easiest way to create a monitoring system using these scripts is to simply call them all from one master script and let each subsidiary script handle its monitoring functions as we've written them. Some alternatives would be to convert our scripts into functions and call them in that manner from a single script or to create a monitoring module and pass various parameters to the module.

The script in Figure III.35 shows one way of calling all of the subsidiary scripts from a master script.

**Figure III.35** master_monitor.pl

```
#master_monitor.pl
do 'get_system_stats.pl';
do 'get_process_stats.pl';
do 'get_port_status.pl';
do 'get_disk_info.pl';
do 'get_db_avail.pl';
do 'get_ws_status.pl';
```

## Master Craftsman

### Scheduling Your Monitoring Scripts

After creating the monitoring scripts, you need to monitor your application. You will then want to automate the execution of those scripts. This has to be done very carefully! If you run your script too frequently, you can cause the script to have a performance impact on the system it is monitoring. If you run it too infrequently, you run the risk of being notified of problems in the system too late to fix them.

A good rule of thumb is to perform a load test and run the scripts at a variety of frequencies. Find the point where the scripts cause minimal to no performance impact, but yet run frequently enough that you can be assured of proactive notification in the case of an error. It may take some time to find the "sweet spot" for executing the monitoring scripts, but it will be worth it when you do.

Of course, there is no error checking in here to make sure that the dependency scripts exist, and so forth. When converting these scripts for your own use, ensure that you add appropriate error checking and management to every script. Using the master_monitor.pl script basically calls each of our monitoring scripts individually and allows them to process their own monitoring routines. You could gain some significant advantages in error checking and ease of use by converting these into functions or even into modules.

# Summary

As previously mentioned, this is only one way that a monitoring system can be written and implemented using Perl. Many other options are available to you, allowing you to expand on the work demonstrated in this chapter to create a monitoring system that fits your specific needs.

Creating monitors for each layer of your application(s), validating the result of those monitors, and then taking appropriate actions are the foundation of any monitoring system. Further enhancements can be done to visually display the results, track historical data, and use trending information to analyze the mean time to failure or other important statistics. All of these enhancements are "icing on the cake" so to speak and rely on the base monitoring foundation to get the data to work with. As always, build a strong foundation and it will support anything that you put on top of it.

# Index

Printed and bound by CPI Group (UK) Ltd, Croydon, CR0 4YY

03/10/2024

01040341-0002